VOLUME SIXTY FIVE

Advances in
MARINE BIOLOGY

ADVANCES IN MARINE BIOLOGY

VOLUME SIXTY FIVE

ADVANCES IN
MARINE BIOLOGY

Edited by

MICHAEL LESSER

Department of Molecular, Cellular and Biomedical Sciences
University of New Hampshire, Durham, USA

ELSEVIER

AMSTERDAM • BOSTON • HEIDELBERG • LONDON
NEW YORK • OXFORD • PARIS • SAN DIEGO
SAN FRANCISCO • SINGAPORE • SYDNEY • TOKYO
Academic Press is an imprint of Elsevier

Academic Press is an imprint of Elsevier
32 Jamestown Road, London NW1 7BY, UK
Radarweg 29, PO Box 211, 1000 AE Amsterdam, The Netherlands
The Boulevard, Langford Lane, Kidlington, Oxford, OX5 1GB, UK
225 Wyman Street, Waltham, MA 02451, USA
525 B Street, Suite 1800, San Diego, CA 92101-4495, USA

First edition 2013

ISBN: 978-0-12-410498-3
ISSN: 0065-2881

For information on all Academic Press publications
visit our website at store.elsevier.com

Printed and bound in UK

13 14 15 16   11 10 9 8 7 6 5 4 3 2 1

Working together
to grow libraries in
developing countries

www.elsevier.com • www.bookaid.org

CONTRIBUTORS TO VOLUME 65

Martin J. Attrill
Marine Institute, School of Marine Science and Engineering, University of Plymouth, Drake Circus, Plymouth, United Kingdom

Isobel S.M. Bloor
Marine Biological Association of the United Kingdom, The Laboratory, Citadel Hill, and Marine Institute, School of Marine Science and Engineering, University of Plymouth, Drake Circus, Plymouth, United Kingdom

Emma L. Jackson
Marine Biological Association of the United Kingdom, The Laboratory, Citadel Hill; Marine Institute, School of Marine Science and Engineering, University of Plymouth, Drake Circus, Plymouth, United Kingdom, and Central Queensland University, School of Medical and Applied Sciences, PO Box 1319, Gladstone, Queensland, Australia

Dianna K. Padilla
Department of Ecology and Evolution, Stony Brook University, Stony Brook, New York, USA

Leif K. Rasmuson
University of Oregon, Oregon Institute of Marine Biology, P.O. BOX 5389, Charleston, Oregon, USA

Monique M. Savedo
Department of Ecology and Evolution, Stony Brook University, Stony Brook, New York, USA

CONTENTS

SERIES CONTENTS FOR LAST FIFTEEN YEARS*

*The full list of contents for volumes 1–37 can be found in volume 38

A Review of the Factors Influencing Spawning, Early Life Stage Survival and Recruitment Variability in the Common Cuttlefish (*Sepia officinalis*)

Isobel S.M. Bloor[*,†,1], Martin J. Attrill[†], Emma L. Jackson[*,†,‡]

[*]Marine Biological Association of the United Kingdom, The Laboratory, Citadel Hill, Plymouth, United Kingdom
[†]Marine Institute, School of Marine Science and Engineering, University of Plymouth, Drake Circus, Plymouth, United Kingdom
[‡]Central Queensland University, School of Medical and Applied Sciences, PO Box 1319, Gladstone, Queensland, Australia
[1]Corresponding author: e-mail address: ismbloor@gmail.com

Contents

Advances in Marine Biology, Volume 65
ISSN 0065-2881
http://dx.doi.org/10.1016/B978-0-12-410498-3.00001-X

Abstract

Global landings of cephalopods (cuttlefish, squid and octopus) have increased dramatically over the past 50 years and now constitute almost 5% of the total world's fisheries
production. At a time when landings of many traditional fin-fish stocks are continuing to
experience a global decline as a result of over-exploitation, it is expected that fishing
pressure on cephalopod stocks will continue to rise as the fishing industry switch their
focus onto these non-quota species. However, long-term trends indicate that landings
may have begun to plateau or even decrease.

In European waters, cuttlefish are among the most important commercial cephalopod resource and are currently the highest yielding cephalopod group harvested in
the north-east Atlantic, with the English Channel supporting the main fishery for this
species. Recruitment variability in this short-lived species drives large fluctuations in
landings. In order to provide sustainable management for *Sepia officinalis* populations,
it is essential that we first have a thorough understanding of the ecology and life history
of this species, in particular, the factors affecting spawning, early life stage (ELS) survival
and recruitment variability.

This review explores how and why such variability exists, starting with the impact of
maternal effects (e.g. navigation, migration and egg laying), moving onto the direct
impact of environmental variation on embryonic and ELSs and culminating on the
impacts that these variations (maternal and environmental) have at a population level
on annual recruitment success. Understanding these factors is critical to the effective
management of expanding fisheries for this species.

Keywords: Cephalopods, Cuttlefish, *Sepia officinalis*, Recruitment, Maternal effects,
English Channel, Spawning, Early life history

1. INTRODUCTION

Global landings of cephalopods (cuttlefish, squid and octopus) have
increased dramatically, rising from 0.5 million tonnes (t) in 1958 (FAO,
1964) to over 4 million t in 2008 (FAO, 2010), and now constitute ~5%
of the total world's fisheries production (FAO, 2010). At a time when landings of many traditional fin-fish stocks are continuing to experience a global
decline as a result of over-exploitation, it is expected that fishing pressure on
cephalopod stocks will continue rising as the fishing industry switch their
focus onto these non-quota species. However, long-term trends may indicate that global cephalopod landings have now begun to plateau or even
show a slight decrease, and a better understanding of these commercial

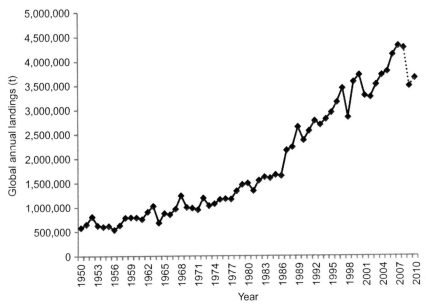

Figure 1.1 A graph showing global landings data for cephalopods (cuttlefish, squid, and octopus) between 1950 and 2010. During the past 5 years (2005–2010), landings have begun to plateau and possibly even decline (2008–2010) (FAO, 2012). *Produced from data in the FAO Global Capture Production 1950–2010 online database.*

cephalopod species is required in order to sustainably manage these stocks (Figure 1.1).

In European waters, cuttlefish are among the most important commercial cephalopod resource (Denis and Robin, 2001; Perez-Losada et al., 1999; Pierce et al., 2010) and the highest yielding cephalopod group harvested in the north-east Atlantic (Royer et al., 2006). Several species of cuttlefish are present in the area, but landings of the common cuttlefish *Sepia officinalis* (Linnaeus, 1958) dominate (Denis and Robin, 2001). The English Channel supports the main fishery for this species (Dunn, 1999; Royer et al., 2006) which is dominated by the French and UK fishing fleets.

1.1. English Channel fishery

Originally considered as a pest species in the United Kingdom due to its low value and copious ink production (Dunn, 1999), landings of cuttlefish by UK vessels have seen a period of rapid increase over the past three decades, rising from approximately 26 t (~£12,000) in 1980 (Dunn, 1999) to almost

4000 t (~£5,500,000) in 2007 (MMO, 2010). The French fishery is better established and landings remain fairly consistent at around 10,000 t a year between 2002 and 2007 (ICES, 2010a). On both sides of the channel, landings are dominated by offshore trawlers (e.g. 90% landings UK) with inshore traps consistently contributing only a minor proportion (e.g. 4–5% landings UK; Denis et al., 2002; ICES, 2003). Inshore landings are highest between March and June (ICES, 2003) coinciding with the peak in breeding season. Offshore landings meanwhile are concentrated in the centre of the channel with a peak between November and March (ICES, 2003), coinciding with the known migration pattern of this population (Boucaud-Camou and Boismery, 1991). Despite the significant increase in exploitation levels, to date, no specific management measures have been introduced in the United Kingdom to maintain and manage the commercially important English Channel cuttlefish stock (e.g. no total allowable catch, no minimum landing size and no fisheries closures; Challier et al., 2005a; Dunn, 1999).

1.2. Stock definition and assessment

The geographic range of *S. officinalis* extends through the Mediterranean Sea and the waters of the Eastern Atlantic to the north-west coast of Africa (Pawson, 1995; Roper et al., 1984). The English Channel represents the northernmost limit for reproduction of this species (Royer et al., 2006). The population here has traditionally been defined and managed as a discrete spawning stock, separate from other populations in its northern distribution, namely, those in the Bay of Biscay and the North Sea (Boucaud-Camou and Boismery, 1991; Boucaud-Camou et al., 1991; Denis and Robin, 2000; Dunn, 1999; Goff and Daguzan, 1991; Pawson, 1995; Royer et al., 2006; Wang et al., 2003).

Stock definition and assessment play an important role in modern fisheries management and are necessary for the development of optimal harvests and quotas, the recognition and protection of spawning and nursery grounds and an understanding of recruitment variability (Begg et al., 1999). Assessment and modelling of *S. officinalis* stocks is difficult not only because of their unique population and life cycle dynamics, which vary from both fin-fish and squid, but also because species distribution and abundance is affected by environmental conditions and fishing is carried out by multiple interacting métiers (Royer et al., 2006). A better understanding of the life cycle, ecology and the effects of environmental fluctuations on recruitment variability is required in order to develop a method for routine stock assessment of *S. officinalis*.

1.3. Life history and life cycle

S. officinalis is a bottom dwelling (nektobenthic) species that occurs on and around the seafloor in habitats with sandy, muddy or rocky substrate (Jereb and Roper, 2005) with or without algal or seagrass beds (Nixon and Mangold, 1998). They are present in both warm and temperate waters at depths of up to 200 m (Boletzky, 1983), after which the ambient pressure is sufficient to implode their internal shell (Ward and Boletzky, 1984). Their exact distribution patterns depend largely on the phase of the life cycle and on ontogenetic migrations (Hanlon and Messenger, 1996). Across its geographic range, this species displays a high degree of life cycle plasticity which allows it to survive in the wide range of environmental conditions encountered (Pierce et al., 2010).

In the English Channel, a consistent biennial life cycle has been described (Boucaud-Camou and Boismery, 1991) which terminates with mass mortality of the adults following spawning (e.g. Rocha et al., 2001). The life cycle (Figure 1.2) begins in spring when mature adult females (Class 2) move

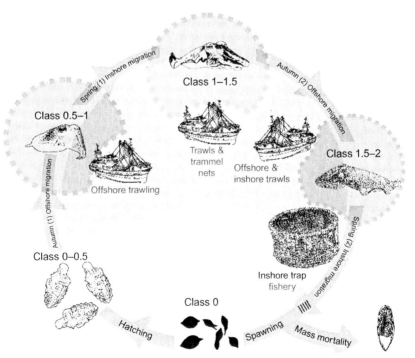

Figure 1.2 Biennial life cycle of *S. officinalis* within the English Channel showing the migration patterns and the main fishery pressures across the life cycle.

inshore to spawn in shallow coastal waters. Hatchlings (Class 0) then emerge during the summer period and undergo a period of rapid growth before beginning their autumn migration offshore to overwintering grounds in the deep central waters of the channel. The following spring juveniles (Class 1) which by now exhibit the first signs of sexual maturation begin their migration back inshore to coastal grounds in search of areas with high food abundance. After a second autumn migration back to their offshore over-wintering grounds, sexually mature adult cuttlefish (Class 2) complete their second and final inshore migration to spawning grounds where they repro-duce and die (Boucaud-Camou and Boismery, 1991; Dunn, 1999; Wang et al., 2003).

1.4. Recruitment variability

In the fisheries context, recruitment is defined as the renewal of a stock by the entry of subsequent young classes to the fishery and determines the num-ber of individuals within a population (Challier et al., 2006). Successful recruitment requires both sufficient breeding rates and subsequent survival of eggs (Boyle, 1990). The short-lived, fast-growing life cycle of S. officinalis means that each year's stock is composed entirely of only one (e.g. Mediter-ranean population) or two (e.g. English Channel population) overlapping generations (e.g. Royer et al., 2006). Such that half (e.g. English Channel) or all (e.g. Mediterranean) of the standing crop of biomass is replaced on an annual basis (Boyle and Boletzky, 1996). The lack of 'buffering' from mul-tiple overlapping generations leaves S. officinalis populations vulnerable to the effects of annual variation in environmental conditions on spawning and early life stage (ELS) survival (Pierce and Guerra, 1994). Annual varia-tions in recruitment success cause large and unpredictable fluctuations in both stock size and landings (Koueta et al., 2000; Piatkowski et al., 2001). Examples of cuttlefish recruitment variability have been demon-strated in both the Gulf of Cadiz (SW Spain), where landings of S. officinalis fluctuated between 505 and 1445 t between 1982 and 1999 (Sobrino et al., 2002), and in the English Channel where variation in annual cuttlefish recruitment was estimated at between 44 and 79 million individuals (Royer et al., 2006).

ELS survival rates are considered to be of particular importance to maintaining recruitment and stock size as following spawning mass mortality of adults is observed (Figure 1.2). Mortality rates are presumed to be at their highest during the ELS (Caddy, 1996), with individuals more sensitive to environmental conditions (e.g. Boyle and Boletzky, 1996; Forsythe,

1993; Rodhouse, 2001). The conditions encountered by pre-recruit ELS (e.g. temperature, oxygen saturation, light, predation and food availability) may account for a significant proportion of the variation in temporal (e.g. between years) and spatial (e.g. between spawning sites) annual recruitment rates. However, other factors such as maternal effects (e.g. oviposition site selection, egg size, egg quality and maternal nutritional status) can also directly or indirectly affect offspring survival and recruitment success. The extent to which these effects subsequently impact recruitment success is dependent on a range of factors including dispersal from spawning grounds.

1.5. Review aims

The main aim of this review is to explore the factors (direct and indirect) that can affect spawning, ELS survival and recruitment variability in *S. officinalis*. This information is essential to enable sustainable fisheries management and a particular focus will be placed on the English Channel which supports the largest commercial fishery for this species. The review has been split into three main sections:

- *Maternal effects*: Explores the effect that maternal life history, phenotype and 'choice' may have on determining variability in an offspring's early life history.
- *Pre-recruit environments*: Examines the effect that heterogeneous environmental conditions (temporal and spatial) within pre-recruit environments have on the early life history.
- *Recruitment to the fishery*: Discusses the potential impact of variation on early life history on the annual recruitment strength of the English Channel population.

Where information on the English Channel *S. officinalis* stock is unavailable; preference was given to information on this species from other parts of Europe or the rest of the World. Where no information concerning this species exists, suitable information from other cephalopod species may also be considered. Important gaps remain in our knowledge of *S. officinalis* ecology and various knowledge gaps have been highlighted and ideas for research to fill these gaps are presented in Section 5.

2. MATERNAL EFFECTS

A maternal effect is defined as the non-genetic effect of the maternal environment or phenotype on the phenotype and performance of offspring (Mousseau and Fox, 1998). Maternal effects act in a variety of ways to enhance (or reduce) offspring fitness (e.g. Donohue, 1999) and have the potential to directly or indirectly contribute to their survival and recruitment success. The role of maternal effects is of great importance to the study of ecology and has been documented in a wide range of terrestrial species (reviewed in Mousseau and Dingle, 1991; Mousseau and Fox, 1998) and more recently in marine species (reviewed in Marshall et al., 2008). Maternal effects are known to affect offspring fitness through their effect on embryonic development (Mousseau and Fox, 1998); these effects can include egg provisioning, propagule size, timing of reproduction, attributes of the offspring's father and spawning location. This review will not take into account paternal effects as for marine systems the data are still considered too few to speculate on this aspect of parental effects (Marshall et al., 2008).

2.1. Mate choice

The extra-genetic contributions of parents to their offspring can vary significantly in quality between different individuals. For this reason, mate choice and sexual selection are discussed and both pre- and post-copulatory mate choice processes are reviewed. For *S. officinalis*, these may include sperm competition (e.g. behavioural aspects see Hanlon et al., 1999), female mate choice (e.g. Boal, 1997) and cryptic female choice (e.g. *Sepia apama*; Naud et al., 2005), with a specific focus on the female aspects of these processes.

2.1.1 Precopulatory female mate choice

Mate choice by females is known to occur in a wide variety of animals including cuttlefish (e.g. *S. officinalis*; Boal, 1997). The classic view of mate choice is that males compete for copulations and fertilisation but that females choose which males to mate with (Bateman, 1948). In this context, female mate choice is considered a maternal effect as the variation in offspring phenotype is a result of maternal behaviour rather than through the transition of maternal genes (Mousseau and Fox, 1998). The cause of this phenotypic offspring variation from mate choice is likely a result of the variation in attributes between different males both genetically and potentially through paternal effects (Mousseau and Fox, 1998).

The concept underlying mate choice is that females mate selectively with males through assessment of some male secondary sexual trait or ornamentation, potentially preferring healthier males or those with good genes. In cephalopods, such secondary sexual traits or ornamentations have not been well defined; however, exhibition of sexually dimorphic body patterns (e.g. Zebra displays) has been suggested as one such mechanism by which females may gauge the health or vigour of competing males (Hanlon and Messenger, 1996). However, Boal (1997) found that while female cuttlefish consistently showed a preference for some males over others, these choices were not based on characteristics known to correlate with male dominance (e.g. body patterning or body size). The only display characteristic that did correlate with female mate preference was the absence of zebra banding, perhaps because the intense zebra display infers agonistic behaviour in nature and thus may actually repel rather than attract females (Boal, 1997). While Boal (1997) demonstrated that under laboratory conditions female cuttlefish play an active role in locating and choosing between potential mates, the existence (and indeed opportunity) for such behaviour in natural populations remains undetermined. Whether cuttlefish choose to mate prior to their migration inshore, enroute to, or at their chosen coastal spawning grounds could have a significant effect on the quantity and quality of mate choices available.

2.1.2 Post-copulatory female mate choice

Copulation is often assumed the final criterion for female choice, but in animals that adopt internal fertilisation strategies, copulation seldom results in direct or inevitable fertilisation (Eberhard, 1985). In some animals, processes after copulation (e.g. sperm competition or cryptic female choice) can still affect the chance of a copulation ending in fertilisation (Eberhard, 1996). The 'cryptic female choice hypothesis' suggests that by actively influencing which sperm is utilised for fertilising eggs after copulation, polyandrous females who have mated with multiple males can still control which male ultimately sires her offspring (Eberhard, 1996).

Female cuttlefish are polyandrous and often accept and store sperm packages from multiple males in the copulatory pouch, located under the buccal mass (Hanlon et al., 1999). It is possible that stored sperm could be manipulated by the females. Naud et al. (2005) demonstrated a potential for female mate choice in S. apama by using microsatellite DNA analysis to determine the genetic diversity of stored sperm with that of subsequent offspring genotypes. S. officinalis adopt a similar polyandrous mating strategy and sperm storage mechanisms and so the potential for post-copulatory female mate choice may also exist for this species.

2.2. Offspring release (oviposition timing)

Where and when offspring become independent are important factors in terms of maternal effects (Resetarits, 1996) and the subsequent impact on recruitment success. In marine environments, habitats can vary significantly (both spatially and temporally) in quality. Natural selection should therefore favour mothers that release offspring at times and locations which increase offspring fitness (Marshall et al., 2008). This 'choice' may be the single most important factor in determining offspring success (Mousseau and Fox, 1998). Two major mechanisms and associated processes of offspring release that can affect offspring fitness and performance are discussed: temporal variation in offspring release or 'oviposition timing' (Section 2.2) and spatial variation in offspring release or 'oviposition site selection' (Section 2.3).

2.2.1 Spawning season

The length of the spawning season exhibits a degree of variability over the geographical distribution range of this species (Goff and Daguzan, 1991) with temperature acting as the regulating factor (Richard, 1971). In warmer waters ($>10\ ^\circ$C during winter), the breeding season can extend all year round (e.g. Ria de Vigo; Guerra and Castro, 1988), with a peak in spawning during spring. In colder water (e.g. English Channel), the onset of the breeding season occurs when water temperatures rise above 12–13 $^\circ$C and usually continues for only a short period (e.g. 2–3 months; Boletzky, 1983). In the English Channel, reproduction (mating and spawning) generally takes place from early spring to midsummer, following the spring inshore migration (Boucaud-Camou and Boismery, 1991; Dunn, 1999; Royer et al., 2006; Wang et al., 2003).

2.2.2 Reproductive strategies

Under laboratory conditions, S. officinalis is known to exhibit a degree of individual variation in reproductive traits. Females have been observed to spawn only once before dying (Boletzky, 1986b), to display a strategy of continuous chronic spawning lasting several weeks (Boletzky, 1987b) or to adopt a strategy of intermittent spawning with long periods of inactivity and eggs being laid up to 4 or 5 months after mating (Boletzky, 1983, 1987b, 1989).

In an attempt to clarify the confusing status of cephalopod reproductive strategies, Rocha et al. (2001) devised five comprehensive and flexible categories. These are based on the type of ovulation, pattern of spawning and the presence or absence of a growth phase between egg-laying events and include:

1. Spawning once (formerly semelparity):
   * Simultaneous terminal spawning
2. Spawning more than once (formerly iteroparity):

- Polycyclic spawning
- Multiple spawning
- Intermittent terminal spawning
- Continuous spawning

Within this structure, S. *officinalis* is categorised as an '*intermittent terminal spawner*' based on general reproductive traits, including: group-synchronous ovulation, mono-cyclic spawning, batched egg laying, absence of somatic growth between spawning events and a single spawning period which occurs during a short time frame at the end of their life cycle (Rocha et al., 2001). Intermittent terminal spawning can result in individuals spawning in multiple events through the entire duration of the breeding season (with offspring encountering different environmental conditions) or during a single event within only a small fraction of the breeding season (with offspring encountering very similar environmental conditions).

The degree to which intermittent terminal spawning may be exploited within natural populations is unknown. Boletzky (1988) states that although the physiology of S. *officinalis* means that long-continued spawning is possible and observed in captivity, there is no evidence that this capacity is exploited within natural populations, only that the potential exists. A study by Laptikhovsky et al. (2003) investigated the spawning strategies of female S. *officinalis* in the Mediterranean. The 'average' spawning female sampled was found to have released around 1000–3000 eggs prior to capture, a number comparable with the fecundity of 'average' captive females that exhibited intermittent spawning (Laptikhovsky et al., 2003). The authors suggest that this is evidence that intermittent spawning may also occur within natural populations.

2.2.3 Migration cues

The term migration is used here to refer to the seasonal movements of a species between regions where conditions are alternately favourable and unfavourable (Dingle and Drake, 2007). A clear annual migration pattern has already been described for the English Channel population of cuttlefish (Boucaud-Camou and Boismery, 1991), which occur in the same general coastal and offshore areas each year (Wang et al., 2003). This pattern is further supported by fisheries capture data from which monthly changes, or spatio-temporal shifts, in the distribution pattern of cuttlefish can also be inferred (e.g. Denis et al., 2002; Dunn, 1999; Wang et al., 2003). During the spring, sexually mature cuttlefish (Class 2; Figure 1.2) migrate from deep offshore overwintering grounds to shallow coastal waters to reproduce (mate and spawn), after which mass mortality is usually observed (Boletzky, 1983). This migratory pattern occurs as a result of the need for spawning adults to deposit their eggs in a habitat in which the ecological and environmental conditions are optimal (spatially and/or temporally) for survival and growth of their offspring (e.g. Dodson, 1997; Pierce et al., 2008). The patterns (timing and extent) of the migration of sexually mature adult cuttlefish to coastal spawning sites along the English Channel may be important in affecting the chances of successful annual recruitment.

Numerous complex and interacting factors (cues) both external (e.g. environmental conditions) and internal (e.g. regulation of genital maturation and reproduction) are likely to be responsible for regulating the initiation of migrations in cephalopods (Boletzky, 1989; Boucaud-Camou and Boismery, 1991; Mangold-Wirz, 1963; Pierce et al., 2010). In the English Channel, the period of optimal environmental conditions in coastal areas for spawning and hatching of *S. officinalis* may vary widely between years. A degree of flexibility in the timing of the spring inshore migration to these areas would be beneficial. This section examines the key cues involved in regulating the timing and onset of this migration.

2.2.3.1 Temperature

Temperature is an important regulating factor in the life cycle of *S. officinalis*. Individuals are known to only remain active above 10 °C and are unable to survive at temperatures of 7 °C or under (Richard, 1971). The effect of temperature on migration is also well documented. In the Adriatic sea, where sea temperatures never fall below 12 °C, seasonal migrations are not initiated in *S. officinalis* populations (Wolfram et al., 2006). In the English Channel, however, where the temperature can get as low as 4 °C in coastal areas (CEFAS, 2010), seasonal inshore and offshore migrations of *S. officinalis* do occur (e.g. Boucaud-Camou and Boismery, 1991).

Sea surface temperature (SST) was found to correlate with the annual migration patterns of *S. officinalis* in the English Channel (Wang et al., 2003). However, although water temperature appears clearly determinative in the winter migrations, its role in the spring migration is less obvious, with cuttlefish leaving the littoral waters while they are still 14 °C and arriving inshore in April when the water temperature averages only 10–11 °C (Boucaud-Camou and Boismery, 1991). This may suggest that other factors such as photoperiod also play a role in determining the onset of these spring migrations (Boucaud-Camou and Boismery, 1991; Boyle and Boletzky, 1996).

2.2.3.2 Sexual maturation

2.2.3.2.1 Photoperiod Photoperiod is defined as day length or 'the period of daily illumination received by an organism' (Concise Oxford Dictionary, 1999) and remains constant between years at any given geographic location. Sexual maturation in *S. officinalis* is under photoperiod control with long dark periods ($\geq$12 h/day; September onwards) stimulating sexual maturation in the female (Koueta et al., 1995; Richard, 1971). In the English Channel, the inshore migration of *S. officinalis* generally occurs shortly after the spring equinox and for adults is clearly dependent upon reproduction, affecting sexually mature animals first (Boucaud-Camou and Boismery, 1991). Photoperiod has a role in entraining (or synchronising) cuttlefish, which have reached the somatic threshold, to undergo the transformation to sexual maturity and thus being primed ready to migrate (Gauvrit et al., 1997; Richard, 1971).

2.2.3.2.2 Temperature In warmer waters, populations of *S. officinalis* generally attain sexual maturity earlier than those in cold water (Boletzky, 1983). In the English Channel, which is at the northern limit of their distributional range, individuals tend to reproduce at the end of their second year (Boucaud-Camou and Boismery, 1991) and are known as Group II Breeders (GIIB). In the Mediterranean and other warm water areas, individuals tend to reproduce at the end of their first year and are known as Group I Breeders (GIB; Gauvrit et al., 1997). In Southern Brittany and the Bay of Biscay, a third reproductive strategy has been described, with individuals reproducing at either 1 (GIB) or 2 (GIIB) years of age (i.e. two biological cycles co-exist; Gauvrit et al., 1997; Goff and Daguzan, 1991). While this pattern is generally accepted, a study by Dunn (1999) did find a small number of males in the English Channel who had reached maturity at around 12 months (GIB), however, no mature females of this age (GIB) were found, suggesting that a specific GIB spawning group does not exist.

The effects of temperature on reproductive growth (e.g. development of gonads) have also been demonstrated by Richard (1966a,b) in *S. officinalis* from the English Channel. Individuals reared at 20 °C attained sexual maturity at the age of 7 months and at an approximate size of 140-mm mantle length (ML), while at 7 months of age individuals raised at 10 °C remained immature and measured only 50-mm ML (Richard, 1966a,b). In cephalopods reared at high temperatures, the gonad attained the stage at which it is receptive to the optic gland hormone more rapidly, but from this stage onwards, temperature has little or no effect on the further development of the gonad (Mangold, 1987). The potential for temperature to affect the age (and size) at which cephalopods spawn, suggests that interannual variations in temperature could cause individuals to mature faster or slower in a given year, as seen by Dunn (1999). The higher water temperatures experienced in 1994 caused some individuals to mature faster, and thus potentially migrating earlier, thereby explaining the unusual presence of GIB males in this study (Dunn, 1999).

2.3. Offspring release (oviposition site selection)

The eggs of *S. officinalis* are laid on structures fixed to the seabed; so, developing embryos remain at the site of spawning for a significant period of time. The location (oviposition site) that mothers 'select' to lay their eggs can therefore dramatically affect offspring performance and fitness by determining the local environmental conditions in which their offspring will develop (Marshall et al., 2008). Bernardo (1996) states that 'even in species with no direct parental care, when and where and how mothers place their offspring is often the single greatest determinant of offspring success'. As resources can be widely distributed both spatially and temporally, it is unlikely that individuals can rely purely on chance to encounter suitable spawning sites (Odling-Smee and Braithwaite, 2003).

The density of eggs laid at a given spawning site may vary spatially and/or temporally, depending on the environmental conditions. For *S. officinalis*, and other benthic spawners, it is likely that the type of substrate (e.g. Chokka squid *Loligo vulgaris reynaudii*; Sauer et al., 1992), biogenic structure (e.g. *S. apama*; Hall and Hanlon, 2002) and the physical characteristics of the water mass (e.g. *L. vulgaris reynaudii*; Roberts and Sauer, 1994) may be instrumental factors in determining the intensity and location of spawning in any given year (Moltschaniwskyj et al., 2003). Females that lay eggs on structures or in habitats where ELS mortality rates are high or in which growth is poor, will leave fewer descendants than females that oviposit in more suitable habitats. For *S. officinalis*, which may spawn only once, it is essential to their individual reproductive success that spawning females have some process by which to assess the quality of spawning sites based on its specific attributes.

The reasons how and why an individual female cuttlefish may navigate to a specific spawning site remains unknown. Possibilities include regional (or natal) philopatry for an area and/or physiologically driven selection, whereby females may actively assess the relative suitability of potential oviposition sites and discriminate among them based on a set of criteria. Little is known about the ability of *S. officinalis* females to discriminate and respond to variation among potential oviposition sites.

2.3.1 Migration patterns

A general annual migration pattern has been described for *S. officinalis* within the English Channel, with individuals occurring in roughly the same broad areas over repeated years (Wang et al., 2003). From early spring to mid-autumn (spawning season), cuttlefish concentrate in the shallow coastal areas on both sides of the channel (Wang et al., 2003). Juveniles begin to migrate to deeper water in the west and central part of the English Channel in late autumn (Wang et al., 2003). A 4-year (1985–1988) study of the migrations of *S. officinalis* in the English Channel was undertaken by Boucaud-Camou and Boismery (1991) using a variety of methods, including floy tagging, which involves the external attachment of small plastic cylindrical 'floy tags', printed with identification and contact details. The authors estimated that the total duration of the migration of *S. officinalis* from offshore wintering grounds to inshore spawning grounds, on the Normandy coast, was approximately 2 weeks (Boucaud-Camou and Boismery, 1991). The patterns of the spring migration for French coast were also inferred from the results of this study.

2.3.1.1 Water circulation and prevailing current

Many marine species are known to migrate in relation to the prevalent currents, utilising them either as a mechanism for transport or as a directional cue. In the Mediterranean, Perez-Losada et al. (2002) investigated the influence of water currents around the Iberian Peninsula on the migrational patterns of *S. officinalis*. In this area, a large scale, strong ocean current (average speed of 40 cm s^{-1}), known as the Almeria-Oran front, flows in a south-easterly direction away from the Spanish coast (Perez-Losada et al., 2002). The authors suggest that this strong prevalent current almost certainly limits the migrational pattern of this population (Perez-Losada et al., 2002).

While it is unlikely that the migrational pattern of *S. officinalis* is affected as dramatically by the prevalent currents in the English Channel, they may have an impact. Wang et al. (2003) investigated the influence of near-surface Atlantic currents which enter the west part of the English Channel. Through a detailed analysis of historical data sets, their results suggest that water circulation at both a local and meso-scale may influence the annual migration patterns of cuttlefish in the English Channel, with the pattern of migration clearly reflecting the pattern of prevalent currents within the English Channel (Wang et al., 2003). This indicates that the pattern of migration of

S. officinalis within the English Channel is influenced, at least in part, by the effects of the prevalent currents in this area.

2.3.1.2 Temperature
Water temperature or more specifically SST may also affect the extent of cuttlefish migrations. Wang et al. (2003) showed that during migration, cuttlefish in the English Channel expand their distribution further north in warmer years and further south in cooler years. The authors suggest that while this indicates a positive correlation between local abundance of *S. officinalis* and SST during the spawning season, it is difficult to determine whether this reflects a causal link or not (Wang et al., 2003).

2.3.2 Navigation and orientation
Franz and Mallot (2000) define navigation as 'the process of determining and maintaining a trajectory towards a goal location'. It is rarely the case that habitats of equal desirability will be available in all geographic directions simultaneously (Lohmann et al., 2008); therefore, a guidance system is key to allow long distance-directed travel, ensuring arrival at an appropriate destination (Dingle, 1996).

Within the English Channel, *S. officinalis* shows directed movement towards coastal areas, appearing in roughly the same areas in repeated years (Boucaud-Camou and Boismery, 1991; Wang et al., 2003). While it is possible that cuttlefish are simply migrating along an increasing SST (or sea bottom temperature) gradient, rather than relying on any navigational capability, Boucaud-Camou and Boismery (1991) state that cuttlefish leave the offshore wintering grounds when the SST is still 14 °C, arrive in coastal areas in April when the SST is still only 10–11 °C, implying at least some additional basic forms of orientation or navigational ability may be required to direct these movements. *S. officinalis* have the ability to perceive visual cues (Alves et al., 2006; Budelmann, 1996; Karson et al., 2003), chemical cues (Boal and Golden, 1999), polarised light (Shashar et al., 1996; Williamson, 1995) and water movements (Budelmann, 1994; Shohet et al., 2006). Spatial information for navigation and orientation could be obtained from such sources during migration.

2.3.3 Site selection
2.3.3.1 Multiple sites
By laying eggs in multiple sites, individuals are able to spread the risks, of hatching success and ELS survival, that are associated with encountering

locally poor site conditions (e.g. increased predation, sub-optimal environmental conditions and limited food availability; Sauer et al., 2000). Although intermittent spawning is known to occur in captivity (Boletzky, 1987b) and potentially in the field (Laptikhovsky et al., 2003), whether *S. officinalis* utilise a single or multiple spawning sites is unknown.

2.3.3.2 Habitat selection

Eggs are usually laid in clusters in shallow water (<40 m) and attached by means of a basal ring to elongated or oblong structures that are fixed to the seabed (Boletzky, 1983). Each egg is approximately 30–40 mm in length, flask-shaped and stained with a coating of ink that is deposited by the female during egg laying (Boletzky, 1983). To create the basal ring, the female uses her tentacles to draw the gelatinous envelope of the egg into a pair of processes which are then wound around the object to which she is fixing the egg, and secured together (Boletzky, 1983). Eggs have been known to be attached to a wide variety of substrata (Table 1.1) including natural substrata such as plants and leaves (e.g. *Zostera marina*; Blanc and Daguzan, 1998), sessile animals (e.g. tubes of polychaetes *Sabella pavonina*; Blanc and Daguzan, 1998), moving animals (e.g. crabs *Latreillopsis bispinosa*; McLay and Guinot, 1997) or previously deposited cuttlefish eggs (e.g. Blanc and Daguzan, 1998; Boletzky, 1983). In addition, where available, artificial structures are also utilised (e.g. fishing pots, ropes and branches; Blanc and Daguzan, 1998). In the laboratory, in cases where an appropriate support (natural or artificial) was unavailable, females were observed to lay their eggs directly on the tank floor (Boletzky et al., 2006).

Spawning grounds tend to occur in shallow areas (between 5 and 60 m deep) with a sandy substrate (with or without pebbles and/or rocks) and associated flora (e.g. seagrass or seaweeds) or structural fauna (e.g. Sabellid worms); in addition, a salinity of 28 or greater and a temperature of between 9.5 and 20 °C is considered optimal (Boucaud-Camou and Boismery, 1991; Mangold-Wirz, 1963; Paulij et al., 1990a). Such conditions or closely similar ones have been cited as necessary for successful embryonic development of *S. officinalis* (Nixon and Mangold, 1998). A spawning ground may also be selected because of good substrate for burying or protection for ELS from predation.

Environmental variables such as temperature, salinity, depth and vegetation are likely to be important as the qualities of a habitat can confer advantages or disadvantages in terms of food availability, shelter and risk of predation. If individuals are unable to select a spawning habitat in which

Table 1.1 Natural structures that have been recorded as used by *S. officinalis* for egg laying

Structure	Location	References
Seahorse (*Hippocampus ramulosus*)	Rio de Vigo	Guerra and González (2011)
Homolid crab (*Latreillopsis bispinosa*)	Philippines	McLay and Guinot (1997)
Peacock worm (*Sabella pavonina*)	Gulf of Morbihan	Blanc (1998) and Bouchaud (1991b)
	English Channel	Bloor (2012)
Feather duster worm (*Spirographis spallanzani*)	Gulf of Morbihan	Blanc (1998)
Hydriod (*Nemertesia* and *Antennularia* spp.)	English Channel	Richard (1971) and Bloor (2012)
Sea fan (*Gorgonian* sp.)	–	Grimpe (1926)
Sponge (*Porifera* sp.)	English Channel	Bloor (2012)
Seagrass (*Zostera marina*)	Gulf of Morbihan	Blanc (1998) and Bouchaud (1991b)
	English Channel	Bloor (2012)
Sugar kelp (*Saccharina latissima*)	Gulf of Morbihan	Blanc (1998) and Bouchaud (1991b)
	English Channel	Richard (1971) and Bloor (2012)
Serrated wrack (*Fucus serratus*)	Gulf of Morbihan	Bouchaud (1991b)
	English Channel	Richard (1971)
Jap weed (*Sargassum muticum*)	Gul.f of Morbihan	Bouchaud (1991b) and Blanc and Daguzan (1998)
	English Channel	Bloor (2012)
Irish moss (*Chondrus crispus*)	English Channel	Bloor (2012)
Clawed fork weed (*Furcellaria lumbricalis*)	English Channel	Bloor (2012)
Brown fan weed (*Dictyota dichotoma*)	Gulf of Morbihan	Blanc (1998) and Bouchaud (1991b)
Desmarest's flattened weed (*Desmarestia ligulata*)	English Channel	Bloor (2012)

Table 1.1 Natural structures that have been recorded as used by *S. officinalis* for egg laying—cont'd

Structure	Location	References
Under tongue weed (*Hypoglossum woodwardii*)	Gulf of Morbihan	Blanc (1998)
Green branched weed (*Cladophora pellucida*)	Gulf of Morbihan	Blanc (1998)
Solier's red string weed (*Soliera chordalis*)	Gulf of Morbihan	Blanc (1998)
	English Channel	Bloor (2012)
Red algae sp. (*Gymnogrongus* sp.)	English Channel	Bloor (2012)
Cleaved wart weed (*Gracilaria multipartita*)	Gulf of Morbihan	Blanc (1998)
Red algae sp. (*Gracilaria* sp.)	Gulf of Morbihan	Blanc (1998)
Podweed (*Halidrys siliquosa*)	Rio de vigo	Guerra and González (2011)
	English Channel	Bloor (2012)
Bootlace weed (*Chorda filum*)	English Channel	Bloor (2012)
Sea beech (*Delesseria sanguinea*)	English Channel	Bloor (2012)

these factors are suitable for embryonic survival, they will fail to produce offspring. The process by which sexually mature female cuttlefish select their spawning habitat is still unknown.

2.4. Offspring provisioning

Beyond the contribution of a female to the when (oviposition timing) and where (oviposition site selection) of spawning, a female can also contribute directly to the initial status of embryonic cuttlefish through the constituents of her eggs. Variation in intra-species egg characteristics is known to occur in cephalopods and can include variation in egg size, biochemical composition, energy content and viability (e.g. Boavida-Portugal et al., 2010; Laptikhovsky et al., 2003; Pecl, 2001; Steer et al., 2004; Sykes et al., 2009). *S. officinalis* have direct-developing offspring with no larval or planktonic phase. Spawning females may thus be more likely to exhibit adaptive plasticity with regard to offspring provisioning and size in reaction to a particular habitat (Marshall et al., 2008).

Offspring provisioning is one of the more obvious types of maternal effect that has been studied and can have far-reaching consequences for the performance and fitness of offspring. Offspring size is known to have pervasive effects throughout the life histories of marine invertebrates and the phenotype of their offspring, and for the most part, larger offspring are considered to have higher performance and fitness than smaller offspring (Marshall et al., 2008). Propagule size is an important life history trait which can be mediated by maternal effects. The position of eggs within the life cycle provides a means to examine both past and future conditions. In terms of the past, the key attributes of an egg, including its size, tend to vary with maternal identity and attributes, which can vary in response to fluctuations in environmental conditions, particularly food availability, experienced during the female's ELS and/or reproductive season and can have significant flow on effects to the subsequent population structure (Kerrigan, 1997). In terms of the future, egg size and quality can also determine initial offspring resources and size, and variation in initial size can be propagated through the individual's life (and into next generation), as well as influencing the likelihood of surviving to a recruitable age (Mousseau and Fox, 1998).

Maternal nutritional history is known to influence both the condition and the size of eggs in at least some species of cephalopods. Steer et al., 2004 found that in the dumpling squid (*Euprymna tasmanica*), when nutritionally stressed, mothers were found to lay smaller and fewer eggs with higher mortality rates. Maternal ration and reproductive history was found to play an important role in the determination of batch fecundity, egg size, hatching success and potentially hatchling competency in cephalopods (Steer et al., 2004). The effects on growth and survival of ELS as a result of the amount and quality of resources allocated maternally to propagules will be addressed elsewhere (Section 3.1.2). Resource competition on ELS growth and survival rates can also be impacted by a female cuttlefish. For example, the number of eggs produced per clutch and the extent to which a resource patch is superparasitised (i.e. eggs laid on the same patch as another female) can both affect offspring phenotype (Mousseau and Fox, 1998).

3. PRE-RECRUIT ENVIRONMENTS

Following spawning, adult cuttlefish exhibit mass mortality (Boletzky, 1983). Newly hatched eggs are left to develop without the benefit of parental care (e.g. Guerra, 2006). The successful development and survival of these eggs to hatching is almost entirely dictated by the conditions encountered (biological and physical) at a spawning ground during embry-onic development as the egg is fixed to structures on the seabed. This emphasises the important role for regulation of the temporal (e.g. timing of reproduction) and spatial (e.g. site and habitat selection) reproductive out-put by adult females in the successful survival and recruitment of the subse-quent generation to the spawning stock. Eggs laid at different times or locations experience different environmental conditions, including temper-ature, salinity, light intensity, photoperiod, oxygen saturation, pollution and predation during their development. Such factors are thought to affect the rate of embryonic survival, development, growth and size at hatching, which may ultimately vary between habitats, locations and months as a result of their variability among sites. These factors can affect not only the life his-tory characteristics of individuals (e.g. survival, fitness, growth, and matura-tion) but also can subsequently impact recruitment to the adult populations (Boyle and Boletzky, 1996; Pierce et al., 2008).

3.1. Local environment pre-hatching

3.1.1 Temperature

S. officinalis spawns in the coastal waters of the English Channel during the spring and summer months. At this time, water temperature is rising and the majority of eggs are expected to hatch during the summer (Boucaud-Camou and Boismery, 1991; Challier et al., 2005a; Dunn, 1999; Royer, 2002; Wang et al., 2003). Water temperature varies significantly between seasons, years, locations and depths, thus eggs laid at different times or at different locations during the spawning period may encounter different and dynamic temperature conditions (Ambrose, 1988). Water temperature is known to directly affect many aspects of ELS marine larvae, including the duration of embryogenesis, yolk utilisation efficiency and the size and weight at hatching (e.g. Pechenik et al., 1990). These physiological responses are due, at least in part, to the controlling effects that temperature has on metab-olism (Vidal et al., 2002). That temperature is a key regulating factor in recruitment of cephalopod populations, as a result of its effects on ELS is an idea supported by numerous authors (e.g. Boyle and Boletzky, 1996; Challier et al., 2005a; Forsythe et al., 2001; Hatfield et al., 2001; Waluda et al., 1999).

The duration of embryonic development in cephalopods is known to be temperature dependent (Figure 1.3) with an inverse relationship between environmental temperature and embryonic development (e.g. Boletzky, 1975, 1987b; Bouchaud and Daguzan, 1989; Caveriviére et al., 1999;

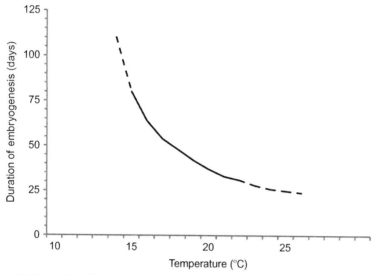

Figure 1.3 A graph to illustrate temperature-dependent embryogenesis in *S. officinalis*. At 20 °C, *S. officinalis* will undergo embryogenesis in 40–45 days while at 15 °C the duration increases to 80–90 days. *Produced from data in Richard (1971) and Boletzky (1983).*

Choe, 1966; Mangold-Wirz, 1963; Palmegiano and d'Apote, 1983; Pascual, 1978). At 20 °C, cuttlefish will undergo embryogenesis in 40–45 days while at 15 °C the duration increases to 80–90 days (Boletzky, 1983). The duration decreases the closer the temperature is to 24 °C and increases as it approaches 15 °C (Bouchaud, 1991b; Figure 1.3). Bouchaud and Daguzan (1990, 1989) demonstrated that embryogenesis can continue at temperatures lower than 15 °C with embryonic development in *S. officinalis* taking 5 months at 12 °C. They also found that while embryonic development ceased at temperatures below 9 °C, when the temperature was increased embryogenesis did restart. This may have implications for spawning in the English Channel suggesting that all year spawning and/or offshore spawning where temperatures would be lower could be possible, although the time frame for development would be much longer.

Bouchaud (1991a) demonstrated that hatching size and yolk volume are also inversely correlated with incubation temperature in *S. officinalis*. Cuttlefish, incubated at higher temperatures, have a lower rate of yolk utilisation, developing faster but also hatching at a smaller size, due in part to the reduced time available for yolk consumption. The embryo thus consumes a smaller fraction of the available yolk prior to hatching than embryos incubated at lower temperatures over longer time periods (Bouchaud and

Daguzan, 1990). *S. officinalis*, hatched at 24 °C, develop faster but are also significantly smaller (DML or dry weight) than those hatched at 15 °C (Boletzky, 1975; Bouchaud and Daguzan, 1989, 1990; Richard, 1971). *S. officinalis* hatchlings are equipped with an inner yolk sac which is utilised in the first 24–48 h of independent life following hatching to prevent starvation (Boletzky, 1987a; Wells, 1958). As well as affecting the rate of embryogenesis and yolk utilisation prior to hatching, water temperature is also thought to influence the production of the inner yolk sac that is essential for post-hatching feeding and survival (Boletzky, 1994). The inner yolk sac provides the main energy source for growth during the first few days after hatching and it is likely that variations in the size of the inner yolk sac, between different sized hatchlings, may have an important effect on their capacity for growth and survival during this time (Boletzky, 1994). A study by Bouchaud (1991a) found that at higher temperatures cuttlefish develop faster, hatch at a smaller size and with a smaller inner yolk sac. These individuals not only have a reduced survival potential due to their smaller size but also have less time following hatching, than those incubated at lower temperatures, that hatch at a larger size and with a greater volume of yolk contained in the inner yolk sac, to find a suitable food source before their yolk reserves run out and they die (Boletzky, 1994; Bouchaud, 1991a). The optimal temperature range for embryonic development in *S. officinalis* could be considered between 15 and 18 °C; within this range, the fraction of yolk utilised for growth becomes highest (Bouchaud, 1991a). While the majority of eggs are considered to hatch during the summer, it is apparent that variations in the dynamic temperature conditions experienced by embryos laid at different spawning sites and times can affect the duration of embryogenesis and subsequently the timing of hatching (e.g. Ambrose, 1988).

In order to maximise survival of ELS, hatchlings must encounter favourable environmental conditions. Warmer temperatures are associated with fast growth and increased food availability. Spawning in the English Channel takes place over a period of several months and so this form of temperature control allows a degree of plasticity that is automatically controlled during embryogenesis as a result of the environmental conditions encountered by individual embryos, both spatially and temporally. Thus, in order for eggs spawned early in the season (e.g. April) to encounter optimal hatching conditions, a longer (e.g. 90 days) embryonic phase allows slower development, producing hatchings which encounter warmer conditions (e.g. July), and the greater volume of yolk contained within the inner yolk sac allows these hatchlings to survive longer after hatching before finding a suitable prey source (Boletzky, 1994; Bouchaud, 1991a). In contrast, eggs laid later in the spawning season (e.g. July) encounter warmer incubation temperatures and develop faster (e.g. 30 days). While this faster development initially seems disadvantageous given that it produces smaller hatchlings with a smaller volume of yolk in the inner yolk sac, indicating a reduced survival potential, when compared with larger, slower developed hatchlings, this disadvantage is quickly compensated for by the hatchlings emerging in favourable environmental conditions (e.g. August) enabling higher growth rates and larger weights to be attained faster than if embryogenesis had been longer and hatchlings had emerged later in the season, to encounter colder waters (Vidal et al., 2002). Thus, temperature may have an influence on the population of juveniles and adults, in relation to both the rate of embryonic development and the conditions experienced as a hatchling and as a result of the quantity of inner yolk reserves, and as such the length of time they have following hatching to find a food source. The variability introduced into embryogenic development as a result of the physiological responses of embryonic metabolism to temperature may help reduce the risk of recruitment failure, by concentrating the bulk of hatching to correspond with the months of optimal environmental conditions for hatching.

3.1.2 Nutrition

In cephalopods, embryogenesis is fuelled by energy reserves stored as yolk within the egg (Sykes et al., 2009). In *S. officinalis*, the latter stage of embryogenic development is characterised by an accumulation of yolk transferred from the outer yolk sac to the inner yolk sac (Boletzky, 1983). The quantity of the yolk in the inner yolk sac is known to vary significantly between individuals and is likely to be affected, at least in part, by the initial size of the egg at laying (Bouchaud, 1991a). It has also been suggested that the quantity of yolk left in the inner yolk sac may influence the maturation of the central nervous system (Dickel et al., 1997). After hatching, the inner yolk sac is used to provide energy during the first 24–48 h of independent life (Bouchaud, 1991a; Wells, 1958), preventing immediate starvation (Boletzky, 1983). A suitable prey source (mysids or similar crustaceans) must be located before the reserves of the inner yolk sac are exhausted (~3 days; Bouchaud, 1991a; Nixon and Mangold, 1998).

The initial provisioning of yolk reserves is controlled by the spawning female and may be affected by their nutritional and thermal history. Using the dumpling squid *E. tasmanica*, Steer et al. (2004) examined the effects of maternal nutrition (high and low rations) and thermal history (summer and

winter) on the quality of eggs produced and the subsequent rate of embryonic survival. Their results indicated that differences in embryonic mortality
and reproductive output were the consequences of maternal ration and not
temperature. Females maintained on low rations laid smaller clutches, with
smaller eggs and higher rates of embryonic mortality when compared to
females maintained on high rations. While the quality, in terms of fatty acid
and lipid content, of the maternally derived yolk was maintained, the quantity of yolk produced for each egg was insufficient to fuel embryogenesis
(Steer et al., 2004). Bouchaud and Galois (1990) also found that in
S. officinalis, the quantity of lipids, which may affect egg quality, varied significantly with egg size. The results of these studies indicate that survival of
ELS can be affected by the maternal nutritional history. The importance of
these findings is that fluctuations in the environmental conditions, especially
food availability, experienced not only during the embryonic and ELS of an
individual but also during the reproductive and ELS of their mother can
impact their life history characteristics such as growth and survival. This is
because the maternally derived yolk is used to fuel the costly process of
embryogenesis, creating a flow of effects which may significantly impact
on the subsequent population structure (Steer et al., 2003, 2004).

3.1.2.1 Food imprinting

The switch from endogenous to exogenous food sources is considered the
most critical period in the life cycle of cephalopods. The lack of parental care
exhibited by *S. officinalis* and the independent nature of the hatchlings means
that information about suitable sources of prey and their availability must be
acquired, by a process independent of parental guidance. Two main processes exist by which cuttlefish hatchlings could acquire information relating
to what and how to obtain suitable prey: an unlearned preference for food or
learning through trial and error (Healy, 2006). A third intermediate option is
'imprinting'. Imprinting has been applied to a wide variety of behaviours,
including habitat preference and refers to a flexible means of learning information for which the timing of the learning is predictable, but the exact
details of the experience are not (e.g. prey preference in hatchlings;
Darmaillacq et al., 2006a; Healy, 2006).

Newly hatched *S. officinalis* are known to innately recognise, prefer and
capture shrimp-like prey during the first weeks post-hatching (Darmaillacq
et al., 2004; Wells, 1958). However, Darmaillacq et al. (2006a) demonstrated that imprinting is evident in cuttlefish hatchlings. Darmaillacq
et al. (2006b) demonstrated that within the first few hours of post-hatching

life, exposure of hatchlings to visual cues of crabs (a naturally non-preferred prey) could be used to override their general preference for shrimp-like prey (e.g. mysids) and instead imprint a preference for crabs, and that the effects of these visual cues overcame those of the first food ingested (Darmaillacq et al., 2006a). Exposure to a visual cue for only 5 h immediately after hatching is sufficient to change their innate preference (Darmaillacq et al., 2006b) and this switch in preference occurs without reinforcement (i.e. no consumption of the prey) and is evident only 3 days later, when hatchlings switch from endogenous to exogenous feeding (Darmaillacq et al., 2006b). This form of learning has been characterised as food imprinting (Darmaillacq et al., 2006a). Since there is a high likelihood of the availability of prey sources to change both spatially and temporally, the flexibility to alter their innate prey preference is likely to be advantageous. Post-hatching visual familiarisation and subsequent food imprinting could allow juveniles to maximise their chances of post-hatching survival by allowing them to modify their feeding preference and thus protecting them from starvation if shrimp-like prey is not available (Healy, 2006).

Darmaillacq et al. (2008) extended their hypothesis of food imprinting to the embryonic phase of cuttlefish development. They showed that by exposing cuttlefish embryos to crabs for a minimum of 1 week prior to hatching, a subsequent visual preference for crabs was observed in ingestively naive 7-day-old juveniles. This work shows for the first time that embryonic visual learning can occur in animals (Darmaillacq et al., 2008). The ability of embryos to learn the visual characteristics of the prey *in ovo* could facilitate post-hatching imprinting on those prey (Darmaillacq et al., 2008). This would allow them to adapt and respond to the individual environmental conditions of their habitat and to learn which prey and predators to hunt and avoid even prior to hatching.

3.1.3 Salinity

The eggs of *S. officinalis* are fixed to a variety of structures on the sea floor, rendering them stationary and subject to any fluctuations in salinity that may occur at the spawning site (e.g. river outflow, rainfall, etc.). Paulij et al. (1990a) investigated the effects of salinity on the embryonic development of *S. officinalis* eggs in the delta. They found that the developmental rate of embryos was significantly reduced at a salinity of below 28.7 and that embryo malformations occurred at a salinity of below 22.4 (Paulij et al., 1990a). Reduced salinity may cause the developing embryo to experience increased osmotic stress which creates a large energy demand and reduces the energy reserves available for successful development (Paulij et al., 1990a). For normal embryonic development to occur, a salinity of 25 or above is generally required (Boletzky, 1983; Paulij et al., 1990a). The highest hatching percentages were found at salinities above 29.8 (Paulij et al., 1990a). However, Boletzky (1983) found that by slowly acclimatising ELS *S. officinalis* to salinity changes, it is possible for them to survive for some time at a salinity as low as 18.

Palmegiano and D'Apote (1983) studied the combined effects of salinity and temperature on embryogenesis and hatching in *S. officinalis*. Their results indicated that, within the temperature range tested (15, 18, and 21 °C), salinity had a significant effect, with cessation of hatching at salinities below 25 (Palmegiano and d'Apote, 1983). Conversely, within this range, temperature had no significant effect on the inhibition of hatching either alone or as an interactive effect between the two parameters (temperature and salinity; Palmegiano and d'Apote, 1983). These results are supported by those of Paulij et al. (1990a) who found that at 17 °C, embryo survival was high at salinities of 26.52 and inhibited at lower salinities.

The effects of salinity on embryogenesis and hatching are important and confirms *S. officinalis* niche as an essentially marine species, unsuited to spawning in brackish water (Palmegiano and d'Apote, 1983). In order to maximise the survival potential of eggs and hatchlings, sexually mature females must select a spawning site with a suitable salinity, in an area where fluctuations (e.g. river input, rainfall runoff, etc.) are minimal.

3.1.4 Oxygen saturation

The dissolved oxygen concentration of the surrounding water is another determining factor in ELS development and survival. In order to obtain optimum rates of development, the oxygen content of the water surrounding *S. officinalis* eggs must be close to saturation (Boletzky, 1983). Limited *in situ* research has been undertaken to investigate how the different oxygen saturations of the surrounding water may affect embryonic survival and development in *S. officinalis* among different spawning locations/habitats, and how this may affect subsequent recruitment and population dynamics.

The egg capsule is designed to protect embryos from predation, but it also creates a range of physiological challenges for the developing embryos, for example, the egg wall provides a barrier to gas diffusion (Gutowska and Melzner, 2009). Oxygen consumption rates significantly increase during embryonic development (Cronin and Seymour, 2000). Both the morphology of the egg capsule and the availability of ambient oxygen in the water are limiting factors in the rate of embryonic oxygen consumption in aquatic environments (Cohen and Strathmann, 1996; Strathmann and Chaffee, 1984; Strathmann and Strathmann, 1995). Water has a low diffusivity and capacitance of oxygen thus limiting its acquisition by developing embryos, in comparison with air. If the developing embryos do not possess a system for active oxygen uptake, then they must obtain oxygen from the surrounding water by diffusion into the egg capsule; this is a slow process and only suitable to supply oxygen in small organisms (Cronin and Seymour, 2000). In order to fulfil this increasing demand, S. officinalis eggs swell during development, thereby reducing the thickness of the egg wall and simultaneously increasing the surface area across which gas diffusion can occur, allowing the embryo to meet their increasing demand for oxygen through diffusion (e.g. S. apama; Cronin and Seymour, 2000). In S. apama, it has also been demonstrated that the developing embryos move within the egg capsule in order to produce convective currents that will prevent the formation of an oxygen gradient within the egg fluid, thereby maintaining the diffusion gradient in order to optimise oxygen diffusion rates (Cronin and Seymour, 2000).

The presence of biofouling organisms on the outside of the egg may also effect the supply of oxygen to the embryo (Cohen and Strathmann, 1996). Unicellular algae and other microorganisms are known to foul a wide variety of egg masses in coastal waters (Cohen and Strathmann, 1996); it is likely that S. officinalis is no exception. These thin films of fouling microorganisms can affect oxygen supply to the embryo through photosynthesis and respiration, as has already been demonstrated in some species of amphibians (Bachmann et al., 1986). For example, photosynthesis by fouling algae on salamander eggs was found to exceed consumption during light phases, while in dark phases, respiration by fouling algae was found to severely deplete oxygen in dark phases (Bachmann et al., 1986; Pinder and Friet, 1994).

In S. apama, it was found that in spite of the combination of egg swelling and the production of internal convection currents through embryonic movement, the internal oxygen concentration continues to decline during embryonic development, from 14 kilopascals (kPa) at the beginning of development to around 6 kPa near to hatching (Cronin and Seymour, 2000). It has been suggested that this reduction in internal oxygen concentration might provide the trigger for hatching once a critical value has been reached. In both S. apama and S. officinalis, this critical oxygen value has been placed at around 5–8 kPa (Cronin and Seymour, 2000).

As outlined above, there are numerous environmental factors that can affect oxygen concentrations and consumption by developing embryos.

An adequate supply of oxygen for successful embryonic development can be maintained through oxygen concentrations in the surrounding water as close to saturation as possible and the presence of photosynthesising micro-organisms on the exterior of the eggs. Conversely, low oxygen concentrations and the presence of respiring microorganisms on the exterior surface of the egg can cause hypoxic conditions for the developing embryo, thereby delaying or arresting embryonic development. Thus, in order for successful embryonic development to occur, spawning cuttlefish must choose a deposition site that is favourable to the conditions influencing oxygen supply to the developing embryos (Cohen and Strathmann, 1996).

3.1.5 Light–dark cycles

Field studies have shown that hatching and subsequent rapid growth in ELS of cuttlefish tend to coincide with the longest day of the year (Koueta and Boucaud-Camou, 2003). In the English Channel, the majority of *S. officinalis* eggs are laid in relatively shallow waters (<40 m; Boletzky, 1983), at which depth daylight is only slightly attenuated, such that approximately 10% of the original intensity of light can still be detected (Poole and Atkins, 1937). Paulij et al. (1991) indicated that in *S. officinalis* embryos, the transition from light to dark acts as a synchroniser for hatching. The same phenomenon has also been described for hatching in *Loligo vulgaris* and *Loligo forbesi* (Paulij et al., 1990b). This process of synchronisation acts to inform organisms about the period and phase of the universal time (Sollberger, 1965). If developing embryos are able to detect transitions between light and dark, then hatching can be synchronised to occur shortly after the onset of darkness (Paulij et al., 1990b, 1991). Light is thought to be an important factor in determining the onset of hatching in *S. officinalis*. The ecological benefits of hatching in the darkness are thought to include a reduction in predation by fish (Paulij et al., 1991).

Unlike the egg capsules of *L. vulgaris* and *L. forbesi*, those of *S. officinalis* are not translucent, but stained black with ink (Paulij et al., 1990b). As embryonic development progresses, the volume of the perivitelline fluid inside the egg increases causing the egg to swell and the black, outer egg capsule to reduce in thickness, becoming more penetrable to light (Paulij et al., 1991). This allows the embryo to be capable of detecting changes in the light cycle during late embryogenesis enabling the hatching rhythm to be entirely driven by external light–dark conditions (Paulij et al., 1991). Paulij et al. (1991) demonstrated that in the absence of a light–dark cycle, a lack of hatching rhythm was evident, with embryos maintained under a continuous light cycle exhibiting an extended time to hatching, and embryos emerged from their capsules as soon as development was complete, irrespective of the time of day (Paulij et al., 1991). In addition, in eggs that were translucent (i.e. where the outer envelopes had not been stained black with ink), the time to hatching was reduced, but the hatching rhythm was maintained. This may be explained by the embryos heightened sensitivity to external light–dark rhythms as a result of the lack of pigmentation in the egg capsule (Paulij et al., 1991).

In addition to the light–dark cycle, the effect of light intensity on the duration of embryogenesis in cephalopods has also been investigated (e.g. Ikeda et al., 2004). Ikeda et al. (2004) found that light intensity did not affect the duration of embryogenesis in the loliginid squid, *Heterololigo bleekeri*. Villanueva et al. (2007) also found that light intensity had no effect on the growth rates of statoliths in paralarval *L. vulgaris*. However, in the cuttlefish species *Sepioteuthis australis*, embryos were found to be affected by light intensity, such that under a regime of summer photoperiod and water temperature, a medium light intensity led to optimal growth of embryonic statoliths (Villanueva et al., 2007).

3.1.6 Water quality

Eggs laid in shallow, coastal waters are often subjected to acute and/or chronic exposure to a variety of contaminants (e.g. trace metals) which are released into the marine environment from agricultural, domestic and industrial anthropogenic activities (Boyle and Boletzky, 1996; Bustamante et al., 2006; Lacoue-Labarthe et al., 2009). *S. officinalis* eggs are attached to fixed structures on the seabed and remain vulnerable to contaminant exposure over the entire duration of embryogenesis and continue on into ELS while individuals remain in coastal waters, until they migrate offshore for the winter (Lacoue-Labarthe et al., 2010).

Our knowledge of the toxic effects of trace elements on cephalopod embryos and ELS is currently limited to a few metals (e.g. Ag, Cd, Co and Zn; Bustamante et al., 2004, 2002, 2006). The uptake of some of these metals (e.g. Zn and Ag) through the outer egg membrane can negatively affect survival and growth of cuttlefish embryos and ELS (Noussithe Koueta personal communication cited in Guerra, 2006). As a result, the water quality at spawning sites, and the degree to which it is contaminated, and with which contaminants, have the potential to impact not only the successful embryogenic development of eggs laid at a site but also, in turn, the subsequent recruitment success and dynamics of the population (Boyle and Boletzky, 1996; Lacoue-Labarthe et al., 2009).

The egg capsule acts as a protective barrier limiting and restricting the incorporation of gases and other chemicals (e.g. trace elements) into the embryo during the first developmental stages (Bustamante et al., 2006; Lacoue-Labarthe et al., 2009). The permeability of the egg capsule is thought to be specific for each element (Bustamante et al., 2002, 2004, 2006; Lacoue-Labarthe et al., 2008b). Many non-essential metals (e.g. Cd, Co, Cs, Pb, Zn or V) are prevented by the egg capsule from being absorbed by the embryo (Bustamante et al., 2002, 2006; Lacoue-Labarthe et al., 2009; Miramand et al., 2006); but nevertheless, the egg capsule is permeable to several toxic elements (e.g. Ag and Mg) which can subsequently be absorbed by the embryo (Bustamante et al., 2004). The permeability of the egg capsule to different elements is not entirely dependent on the metabolic needs of the embryo, since the non-biologically essential element Ag is well known for its enhanced embryotoxicity (Calabrese and Nelson, 1974; Warnau et al., 1996), but rather on the physico-chemical properties of the elements which dictate their capability to pass through the eggshell (Bustamante et al., 2006). The degree of permeability may also change during the course of embryogenesis. As the volume of fluid in the egg increases, the diffusion and retention properties of the egg capsule is also known to change (Cronin and Seymour, 2000; Lacoue-Labarthe et al., 2009).

Following an analysis of the current research on the uptake of trace elements and contaminants by cuttlefish embryos, Bustamante et al. (2006) found that the egg capsule has three main pathways to deal with contaminants present in the ambient water:

1. Adsorption onto the membrane of the egg capsule, shielding the embryo from direct exposure (e.g. ^{241}Am, ^{109}Cd, and ^{57}Co).

2. Adsorption onto the membrane of the egg capsule, providing temporary protection for the embryo; however, once both the specific and non-specific binding sites are filled, these contaminants are able to permeate across the membrane and be absorbed by the embryo (e.g. ^{110}mAg and ^{65}Zn).

3. Permeate the egg capsule membrane in both directions, allowing the embryo direct access to absorb them (e.g. ^{134}Cs).

While Bustamante et al. (2006) have managed to categorise the mechanisms by which contaminants are dealt with by the egg capsule membrane, the specific mechanisms and processes through which the element selectivity of the eggshell is determined (or controlled) remains poorly understood (Bustamante et al., 2006).

Lacoue-Labarthe et al. (2008a) investigated the possibility of a second pathway for embryonic contamination, from maternal transfer of metals to eggs during the pre-spawning period. Gravid females use part of their somatic tissue to produce eggs (Guerra and Castro, 1994), and a small fraction of the contaminants present in an adult female can be transferred to the offspring during this process (Lacoue-Labarthe et al., 2008a). Of the eight elements tested, only two essential elements (Zn and Se) and one non-essential element (Ag) were found to effectively transfer to the vitellus of the eggs produced (Lacoue-Labarthe et al., 2008a). While Se and Zn are essential elements well known for their maternal transfer (Tsui and Wang, 2007; Unrine et al., 2006), Ag has no apparent metabolic role and its transfer to eggs may thus provide a mechanism for maternal depuration (Lacoue-Labarthe et al., 2008a).

There remains little information on the related effects of acute or chronic exposure to trace elements on the development of ELS cuttlefish (e.g. Lacoue-Labarthe et al., 2008a,b). Such work is of interest for several key *S. officinalis* spawning areas in the English Channel (e.g. Bay of Seine) which are known to be high in pollutants (e.g. Zn, Ag, Cd; Calabrese and Nelson, 1974; Lacoue-Labarthe et al., 2009). A better understanding of the effects of contaminant enrichment of these coastal waters on the embryonic and ELS development of *S. officinalis* is essential for assessing its impacts on recruitment and the eventual dynamics of the English Channel cuttlefish population.

3.1.7 Predation

Cuttlefish are susceptible to predation at almost all stages of their life cycle from hatching to spawning; no major predation pressure on the eggs has been reported (Guerra, 2006). Cuttlefish undergo direct embryonic development within well-protected, benthic, sedentary egg capsules; thus, although they receive no parental care, they are expected to be less vulnerable to predation than other types of eggs (e.g. pelagically spawned eggs; Staver and Strathmann, 2002). The cuttlefish egg at up to 15 mm × 30 mm is one of the largest cephalopod eggs (Boletzky, 1983). With the embryo encapsulated inside the egg by three separate layers, this large and sturdy egg capsule provides protection against potential predators (Gutowska and Melzner, 2009) and from microbial infection (Boletzky, 1986a). In addition, to the protection provided by the tough membranes of the egg capsule (e.g. Gutowska and Melzner, 2009), the outer envelope of the egg is stained with black ink. It has been suggested that this black colouration is a strategy to aid in the protection of the embryo through enhanced camouflage against the surrounding environment (Guerra and

González, 2011) and also by potentially acting as a chemical deterrent to predators due to the presence of millimolar levels of free amino acids and ammonium in these secretions, which may cause sensory disruption to potential predators (Derby et al., 2007). Recent evidence from Guerra and González (2011) indicated that predation of eggs in natural environments does still occur with evidence of predation by the Tompot blenny (*Parablennius gattorugine*) on *S. officinalis* eggs in the Ria de Vigo. However, a review of mortality rates in marine embryos by Strathmann (2007) suggests that those in protected benthic aggregations (e.g. *S. officinalis*) are substantially safer (<0.1 d^{-1}) than pelagic embryos (>0.1 d^{-1}; Lamare and Barker, 1999; Rumrill, 1990; Strathmann, 1985).

Cuttlefish do not exhibit a true larval phase, but instead hatch as structurally and behaviourally adept paralarvae that are miniature replicas of the adults (e.g. Hanlon and Messenger, 1988). This lack of planktonic phase effectively reduces the risks associated with ELS and hatching, resulting in increased survivorship when contrasted with many marine species (e.g. squid; Caddy, 1983). However, ELS and juvenile *S. officinalis* are known to be preyed upon by a number of vertebrate predators (e.g. Hastie et al., 2009). In the English Channel, potential predators of juvenile *S. officinalis* include juvenile sea bass (*Dicentrarchus labrax*), juvenile dogfish (*Scyliorhinus canicula*), velvet swimming crabs (*Necora puber*; Langridge et al., 2007) and the smooth hound (*Mustelus mustelus*; Morte et al., 1997). A study by Langridge et al. (2007) demonstrated that juvenile cuttlefish reacted differently to different types of predators through the use of complex discrimination and selective expression of defences. Juvenile *S. officinalis* were found to express deimatic display only towards visually orientated teleost predators and never to more chemosensory predators such as dogfish or crabs (Langridge et al., 2007).

Not only are cephalopod hatchlings vulnerable to predation by fish and other species but also they are known to exhibit cannibalistic traits throughout most life stages, especially when food is limited (e.g. Ibáñez and Keyl, 2010). Intra-cohort cannibalism has been observed in *Sepioteuthis* hatchlings, with larger hatchlings known to readily attack and consume smaller conspecifics (Walsh et al., 2002). The risk of mortality for hatchlings as a result of cannibalism is likely to vary throughout the spawning season increasing at times, or in areas, where egg density is high or as a result of temporal or spatial shifts in prey availability (Steer et al., 2003). In such circumstances where cannibalism is evident, it is likely that size-selective mortality occurs with larger hatchlings achieving higher survival and recruitment rates in comparison to their smaller counterparts.

S. *officinalis* has a soft unarmoured body which means they have little structural defence and instead rely heavily on behavioural responses to avoid predation (Hanlon and Messenger, 1996; Messenger, 2001; Poirier et al., 2004). S. *officinalis* hatchlings have two general defence mechanisms to reduce the risk of predation: primary defences such as crypsis serve to reduce the risk of detection by potential predators, while secondary defences such as inking which distract the predator while they escape (Boyle and Rodhouse, 2005) are only used once the individual has been detected (Hanlon and Messenger, 1996; Messenger, 2001; Poirier et al., 2004). Cuttlefish hatchlings are able to adapt their colouring and texture to reduce their visibility in a wide variety of substrates and habitats. In the natural environment, some substrates and habitats may enable them to camouflage themselves to a better degree. This may be an important factor in determining site selection, or in the survival (and subsequent recruitment) rates of hatchlings from different habitats or locations. For example, in sandy substrates, S. *officinalis* will not only use crypsis but also will partly bury themselves in the substrate to aid their concealment from predators, a tactic which is obviously not possible in rocky habitats (Boletzky, 1987c; Hanlon and Messenger, 1996; Poirier et al., 2004). Little is known about the movements of ELS S. *officinalis* following hatchling; if they remain in their spawning grounds until migrating offshore, then the maternal choice of habitat for egg deposition will greatly affect the efficiency of hatchlings' ability to evade predation through adequate cryptic camouflage or burying. However, if ELS leaves the spawning grounds following hatching, because of the minimal level of predation during embryonic development, the choice of spawning site is less important as hatchlings can simply leave the area in search of habitats suitable for crypsis.

3.1.8 Storm events

In Australia, a study of S. *australis* spawning grounds indicated that a loss of egg masses was sometimes correlated with storm events, on one occasion a loss of 54% of egg masses was recorded at one of the study sites (>10 m water depth), coinciding with a period of strong winds (Moltschaniwskyj and Pecl, 2003). Storm-associated wave action may have been responsible for the egg masses being dislodged at the site (Moltschaniwskyj and Pecl, 2003). Given the shallow nature of the coastal spawning grounds of S. *officinalis* within the English Channel, similar losses may occur as a result of severe storm events, and egg clusters are often noted washed up on beaches following storm events. This raises the importance of exposure (wind and/or wave) as a potential factor in site selection by spawning females, as egg clusters laid at exposed sites risk being dislodged during adverse weather conditions with the potential for high mortality rates as a result.

3.1.9 Eggs on pots

In Morbhihan Bay, Bay of Biscay, Bouchaud (1991b) estimated the total number of cuttlefish eggs laid on cuttlefish traps to be between 18 and 40 million eggs during a single spawning season (March–September). In this area, fishermen haul their traps every 2 days and prior to redeployment, the traps are often stacked and cleaned to remove any eggs, reducing additional weight and drag that the presence of eggs can cause during hauling (Blanc and Daguzan, 1998). There are currently no estimates available for the quantity of cuttlefish eggs laid on cuttlefish traps each season in the English Channel, or for the proportion of these eggs that are lost. Further research is needed to obtain such estimates and to assess what effect the loss of eggs from this source might be having on annual recruitment success of S. officinalis within the English Channel.

3.2. Local environment post-hatching

One of the fundamental assumptions of life history theory is that there is a positive relationship between offspring size and fitness (Marshall et al., 2003). The hypothesis that fast growth enhances the survival rate of young fish is a central tenet in the study of recruitment determination in marine fish (Meekan and Fortier, 1996). Cephalopod growth is rapid and exhibits a high degree of plasticity, creating considerable variation in the intra-specific size-at-age data observed within many species (Arkhipkin and Laptikhovsky, 1994; Jackson, 2004; Jackson and Moltschaniwskyj, 2002) including S. officinalis. Endogenous factors include genetics, maternal contributions, and an individual's previous growth history are all considered to have some impact on growth rates of individuals. In order to understand variability in growth rates, it is also important to know how different exogenous (biotic and abiotic) factors impact on it. In a review by Forsythe and Van Heukelem (1987) the authors concluded that of the 12 biotic and abiotic factors discussed, water temperature and food availability (quantity/quality) were the most important for growth in cephalopods. This is a viewpoint which numerous field and laboratory studies have supported (Pecl, 2004; Pecl et al., 2004; Villanueva et al., 2007).

The majority of cephalopods have a two-phase growth pattern (Forsythe and Van Heukelem, 1987; Wells and Wells, 1970), with a first, exponential growth phase during the ELS (around 60–90 days; Koueta et al., 2000), and a second slower, power growth phase during later life (Forsythe and Van Heukelem, 1987). Exponential growth rates in ELS *S. officinalis* have been demonstrated by several authors (Domingues et al., 2001b; Forsythe et al., 2002; Gutowska and Melzner, 2009), with example growth rates from one laboratory study estimated at between 10.8% and 12.8% body weight per day (BW d^{-1}), for individuals reared at between 19.5 and 20.5 °C during the 40 days after hatching (Koueta and Boucaud-Camou, 2001). This exponential growth phase during the ELS is important as even very small differences in growth rates during this phase can lead to substantial variations in the final adult size of individuals (Forsythe, 1993).

In many cephalopod species, hatchlings are known to emerge over extended temporal (e.g. spring or summer) and spatial scales (e.g. different spawning sites), depending on when and where spawning occurred. This creates a potential for ELS to experience varying environmental conditions, such as temperature and food availability (quantity/quality), both temporally and spatially which can affect individual growth rates, creating the intrinsic variability in size-at-age that is evident within a population and species (Forsythe, 1993, 2004; Forsythe and Hanlon, 1988, 1989; Hatfield, 2000; Jackson et al., 1997). Individual growth rates can also affect the dynamics of the population, through their impact on survival (and subsequent recruitment), with the potential for increased survival rates in faster growing individuals due to their ability to swim faster, escape predators and capture food better. In addition, growth rates can also affect fecundity upon maturation, with faster growing individuals attaining sexual maturation sooner and with the potential to be more fecund as fecundity generally increases with size (e.g. fish, Fuiman, 2002).

3.2.1 Temperature

During the exponential growth phase of ELS cephalopods, temperature can have an immediate and significant effect on growth rate, creating significant variation in eventual adult size. Cephalopods have a poikilothermic metabolism, the rate of which rises or falls directly with temperature, creating a corresponding rise or fall in feeding and growth rates (Forsythe, 1993). In the English Channel, *S. officinalis* spawn at a time (spring to summer) or place (shallow coastal waters) where hatchlings will encounter continually warming temperatures during their first months, thereby optimising the effect that temperature can exert on the exponential growth rates of these ELS (Forsythe, 1993).

The potential impact of this effect on ELS cephalopods was first demonstrated by Forsythe (1993) who hypothesised that, as hatching occurs over a period of gradually warming water temperatures, each monthly cohort will encounter warmer water temperatures during their exponential growth phase and thus grow significantly faster than cohorts that hatched only weeks earlier, enabling younger cohorts to surpass older cohorts in size, prior to reaching 1 year of age. Forsythe (1993) conducted laboratory investigations (using the squid species *L. forbesi*) to investigate this hypothesis. The results showed that during the exponential growth phase, small changes in temperature (1 °C per month) had the potential to produce significant changes in growth (1% BW d^{-1}; Forsythe, 1993). By creating a mathematical model and running simulations based on this data, Forsythe (1993) was able to demonstrate that a 1 °C increase in water temperature could create a twofold increase in body weight over a 90-day period and a 2 °C increase a fivefold increase in body weight; these temperature increases mimic that evident in the coastal waters of the English Channel during the spring/summer hatching of *S. officinalis*. These results indicate that temperature can modulate rapid growth rates during the ELS, when individuals are small and most vulnerable to predation. It is also likely that any enhancement or diminishment of growth rate during this exponential phase could also affect their final adult size, having additional implications for population structure and recruitment rates (Forsythe, 1993). This hypothesis is now well supported with evidence from field and laboratory studies (e.g. Forsythe and Hanlon, 1988; Hatfield, 2000) and is now referred to as the 'Forsythe effect'.

While the Forsythe effect explains part of the variability in growth rate observed in ELS cephalopods, it by no means explains it all. The majority of laboratory experiments investigating the Forsythe effect on ELS cephalopods have been conducted under conditions of consistent and abundant food supply. This situation does not mimic the experience of natural populations. Food availability is another factor that requires consideration when investigating the factors involved in growth rate variability of ELS cephalopods.

3.2.2 Food availability

Cuttlefish hatchlings are predatory, hunting and feeding on live prey including small crustaceans, especially mysids, gammarids and carangids (Boletzky, 1989; DeRusha et al., 1989; Wells, 1958). They are voracious feeders capable of maintaining high feeding rates of approximately 20–30% of their BW d^{-1} (Boucher-Rodoni et al., 1987). Their ability to convert food intake at a high rate enables them to maintain the rapid growth rates often observed in cephalopod species (Boucher-Rodoni et al., 1987). While temperature is often cited as the major factor responsible for differences in growth rate, the availability of food is also a determining factor, as without it, there would be insufficient energy to fuel the increased growth rates observed at higher temperatures (Forsythe and Van Heukelem, 1987). Cuttlefish hatchlings require a minimum quantity of food for survival (e.g. sufficient to fuel metabolism and repair) and additional energy is required if growth is to occur. Thus, one of the factors potentially limiting the growth rate of an individual is the portion of excess food available to fuel growth (Forsythe and Van Heukelem, 1987).

Food availability (i.e. the quantity and quality of food available) is potentially a limiting factor for growth in ELS cephalopods. Laboratory studies have been undertaken investigating the effects of both quantity and quality of food type on growth (e.g. Correia et al., 2008a,b; DeRusha et al., 1989; Domingues et al., 2001a,b, 2003, 2004; Koueta and Boucaud-Camou, 1999, 2001, 2003; Koueta et al., 2000, 2002; Pascual, 1978; Richard, 1971).

The food conversion rate of cephalopods is high (Hanlon and Messenger, 1996; Pascual, 1978; Segawa, 1990), allowing the rapid growth evident in many of these species, with approximately 30–60% of food intake estimated to be used for growth (Koueta and Boucaud-Camou, 2001). The quantity of food available and ingested has been shown to have a significant effect on the growth rates of ELS cephalopods. During the first month of their life, ELS *S. officinalis* require large food rations, with food requirements progressively reducing after this period (Koueta and Boucaud-Camou, 1999), with an optimum ration of 16.2% of their body weight during the first 10 days of rearing and decreasing progressively to 10% of their body weight by day 40 of rearing (Koueta and Boucaud-Camou, 2001). The 'maintenance ration' is the minimum quantity of food required for maintaining growth and is estimated at between 2.5% and 3% of an individual's body weight, below this amount an individual loses weight (Koueta and Boucaud-Camou, 1999). It has also been shown that the quantity of food offered also affects the quantity of food ingested, such that as more food is offered, more food is also ingested, especially during the first 10–20 days of rearing; consequently, increased growth rates were also associated with higher food ingestion (Koueta and Boucaud-Camou, 1999).

The quality of the food source has also been shown to influence growth rates in ELS of *S. officinalis* (Forsythe and Van Heukelem, 1987). Cephalopods are known to have a mainly protein metabolism (Lee, 1994); however, lipid and fatty acid profiles are also considered important for the optimal development of ELS cephalopods (Domingues et al., 2003). In natural populations, the diet of ELS *S. officinalis* consists mainly of crustaceans

(89%), with fish representing a minority intake (4.6%; Blanc et al., 1998). This field data is supported by laboratory studies, with several authors reporting reduced growth rates for ELS cuttlefish when fed frozen or live fish when compared to those fed crustaceans (DeRusha et al., 1989; Pascual, 1978), indicating that fish are not an adequate prey for the culture of ELS cuttlefish, although they do appear to become a more important component of cuttlefish diets at later stages of their life cycles. The results of most prey studies have indicated that for ELS *S. officinalis*, crustaceans are the optimal food for development and growth. The differences in growth rates that are apparent in the growth of ELS fed different diets (e.g. frozen prey, pellets, fish fry) may be associated with the biochemical (e.g. proteins and lipids) composition of these food items, as suggested by Domingues et al. (2003). When crustaceans are the main prey source, growth rates in laboratory studies are generally reported to range between 5% and 9% BW d^{-1} (DeRusha et al., 1989; Domingues et al., 2004; Koueta and Boucaud-Camou, 1999; Koueta et al., 2002; Pascual, 1978) at temperatures around 19–22 °C. However, even within crustacea, different prey items can have a significant effect on growth rates. In a study by Baeza-Rojano et al. (2010), the use of caprellids and gammarids as potential alternatives to mysids was investigated. Caprellid amphipods are small crustaceans which inhabit littoral zones and are found on microalgae, seagrasses, bryozoans and erect hydroids (Guerra-Garcìa and Tierno de Figueroa, 2009); however, the results of their studies indicated a significantly reduced growth rate in those ELS *S. officinalis* fed on caprellids *Caprella equilibra* (1.6 ± 0.25 BW d^{-1}) when compared to those fed mysids and gammarids (>5% BW d^{-1}; Baeza-Rojano et al., 2010).

The implications for natural populations are that the location or timing of spawning could impact ELS growth rates and survival as a result of differences in faunal distribution, causing localised changes in food availability. Mysid, for example, is a key prey item for ELS cuttlefish. Seasonal changes in mysid density occur within the English Channel (e.g. Zouhiri et al., 1998). Thus, cuttlefish hatching in certain months or weeks may be exposed to higher densities of prey than those hatching at other times. As suggested by Koueta and Boucaud-Camou (1999), the exposure of ELS to increased levels of food leads to increased food ingestion and subsequently increased growth, thus creating a potential for variation in growth rates, independent of temperature, between individuals hatching at different points in the season. This concept can also be extended to different spawning site locations, both in terms of their geographical location and also their habitat composition in terms of suitability for different faunal species. For example, Zouhiri et al. (1998) found that the western part of the English Channel was more diversified in terms of mysid species, indicating an impoverishment of species from west to east, such that while ELS hatching in habitats or sites with low levels of crustaceans, could still survive by feeding on other prey items, their reduced quality in terms of biochemical composition, may cause growth rates in these individuals would be reduced, potentially leaving them susceptible to increased predation as a result of the increased time spent in this small vulnerable life phase. This is an important point to consider in terms of the level of oviposition site discrimination and selection by adult females.

3.2.3 Size-at-hatching

While temperature and food availability have often been cited as the two most important factors affecting growth in ELS cephalopods (e.g. Forsythe and Van Heukelem, 1987), it cannot be the only source of variation as observations from experimental rearings indicate that even under identical conditions (e.g. food and temperature), siblings continue to display variation in size-at-age following a period of growth (Forsythe, 1993). Many of these growth studies, including those in support of the Forsythe effect have been based on the assumption that the population is composed of identically sized hatchlings, with growth initiating from a common starting point. However, many cephalopod species, including *S. officinalis*, are known to exhibit a considerable range in size at hatching. Hatchlings of *S. officinalis* are known to range in weight from 0.053 g to at least 0.180 g (Domingues et al., 2001a) and in length from 6 to 9 mm (Boletzky, 1983). Steer et al. (2004) suggest that the origin of such variation in cephalopods size at hatching can generally be classified into three or four areas of origin:

1. *Broad scale environmental effects*: for example, temperature, which can affect the duration of embryogenesis, as well as the production and utilisation of yolk;
2. *Small-scale environmental effects related to the microenvironment of the egg mass*: for example, eggs from large clutches may develop slower than those from smaller clutches (e.g. squid species (*S. australis*); Steer et al., 2003);
3. *Maternal effects*: for example, maternal conditions during oogenesis can have dramatic effects on resource allocation and subsequent hatching condition, with nutritionally stressed individuals potentially producing fewer and smaller eggs, that result in smaller hatchlings (e.g. dumpling squid (*E. tasmanica*); Steer et al., 2004).
4. *Genetic differences*: for example, variation in maternity and paternity.

It is now widely accepted that size-at-hatching may represent an additional factor affecting ELS growth.

In an extension to the Forsythe effect, Pecl et al. (2004) suggested that variability in hatchling size could, theoretically, result in larger hatchlings reaching twice the size of their smaller counter parts after a period of only 90 days growth, assuming that hatchlings of different sizes grow at the same rate when exposed to similar environmental conditions. In addition, the authors hypothesise that contrary to this assumption, a definable relationship between growth rate and hatchling size may exist, with smaller hatchlings experiencing a faster growth rate than their larger counterparts or vice versa (Pecl et al., 2004). This hypothesis was further explored by Leporati et al. (2007), using hatchlings of the octopus species *Octopus pallidus*. Using two treatment temperature ranges (spring/summer and summer/autumn), the authors' aim was to see if different size hatchlings grow at similar rates. The results of this study indicated that at lower temperatures, small hatchlings will grow more slowly (ca 24%) than their larger counterparts, thereby reaching a smaller final size (Leporati et al., 2007). However, at higher temperatures, the authors found that this pattern of growth was displaced, with smaller hatchlings growing faster (~8%) than their larger counterparts, thereby either reaching a similar final size, or even surpassing them in final size (Leporati et al., 2007). The authors suggest that the different relationships between growth rates, hatching size and temperature regime could contribute to the substantial intrinsic intra-specific size-at-age data observed in many cephalopod populations (Leporati et al., 2007). A complex relationship exists between hatching size and growth rate, but this appears to be secondary to the effects of temperature (Leporati et al., 2007). Further investigations are required to determine the true effect of size-at-hatching on ELS growth rates of *S. officinalis*.

3.2.4 Survival

Size at hatching can provide a useful indication of hatchling competency and survival (Pepin, 1988). This phenomenon is often referred to as size-selective mortality, or the 'bigger is better' hypothesis (Steer et al., 2003). It is thought that increased competency in larger hatchlings is the result of reduced sensitivity to starvation and predation, in comparison to smaller hatchlings, due to enhanced foraging success and improved swimming abilities (e.g. Pepin, 1988), with slower growing and smaller individuals within a cohort being preferentially removed (Meekan et al., 2006). It is possible to classify the origin of variation in hatchling size into three main areas: broad-scale environmental conditions (e.g. temperature), small-scale environmental conditions related to the micro-environment of the spawning ground and maternal effects (e.g. nutritional regime; Steer et al., 2004).

Size-selective mortality may be more evident alongside high larval growth rates, which accentuate the size ranges of individuals within a cohort. In addition, in fast-growing species like cephalopods, this size variation will be present at younger ages so that the onset of selection can occur earlier (Meekan et al., 2006). Size selection has been invoked to explain selective mortality in fast-growing temperature species such as the calamari squid (*S. australis*; Steer et al., 2003). This study investigated whether bigger hatchlings have a higher survival and recruitment rate in comparison to smaller hatchlings (Steer et al., 2003). The distribution of Natal Radius (NR) size, an indication of hatchling size, of adult statoliths was skewed further to the right than the hatchling NR size distribution (Steer et al., 2003). This indicates that bigger hatchlings are more likely to survive into adulthood perhaps as a result of size-selective predation mortality during the ELS of *S. australis*, resulting in smaller adults being less likely to recruit to the population (Steer et al., 2003). However, there is potential for individuals to grow out of one window of vulnerability for a particular type of predator and enter into another, suggesting that there may be continuous non-random predation, with animals always running the gauntlet.

However, a purely size-based predation model would oversimplify the process of connectivity that is apparent in the marine environment as there are many other factors involved (Cowan and Shaw, 2002). The occurrence of extreme environmental conditions or events, food availability, fishing pressure, disease and parasites, deteriorating or lost habitat and local hydrodynamics may all be involved in increased mortality (Sogard, 1997), with several demonstrating an element of size selectivity (Steer et al., 2003). Temporal and spatial mismatch between hatchlings and prey sources may also affect ELS survival rates. In many fish species, it is apparent that the risk of starvation decreases during the course of development, partly due to access to increased bodily energy reserves but also as a result of heightened sensory and swimming performance. Older (or larger) hatchlings may be more able to successfully locate distant food supplies, and due to their larger size, they may also have access to a wider range of prey species, such that an increase in diet breadth may be critical for growth rates in ELS of marine species (Fuiman, 2002).

In 1914, Hjort was the first to explicitly link feeding to ELS survival and recruitment to the abundance of food available during the critical transition period of ELS from endogenous (yolk) to exogenous (prey) feeding. The 'Critical Period Hypothesis' stated that should food be limited during this critical transition period, increased mortality of ELS will occur as a result of starvation, but conversely, when food abundance was high, survival of ELS would be high. Hjort further postulated that such variations in survival of ELS as a result of starvation could generate recruitment variability (Hjort, 1914 cited in Fuiman, 2002). The availability and abundance of prey items within different spawning grounds may be a critical determinant in the relative recruitment of cuttlefish to the annual recruitment of the stock in the English Channel. Around 60 years following the Critical Period Hypothesis, David Cushing, an English fishery scientist, extended Hjort's hypothesis in his match/mismatch hypothesis. This stated that food limitation during any of the phases of ELS development, and not just the transitional period from endogenous to exogenous feeding, could provide a major contribution to recruitment variability through its impact on ELS mortality rates. In order to increase the rate of ELS survival, it is essential to match (both spatially and temporally) ELS abundance with seasonal food production (Fuiman, 2002).

4. RECRUITMENT TO THE FISHERY

4.1. Recruitment variability

In the fisheries context, recruitment is defined as the renewal of harvestable stages in a population and is an essential parameter of stock dynamics (Challier et al., 2002). In terms of the English Channel fishery for *S. officinalis*, this refers to the renewal of the stock by the smallest commercial category (i.e. for France BW < 100 g), and is represented by their appearance in the fishery. Landings from the French fishery indicate that, for this smallest commercial category, recruitment in the English Channel begins in the autumn, following their offshore migration (Boucaud-Camou and Boismery, 1991; Challier et al., 2002; Dunn, 1999; Royer, 2002).

Annual recruitment variability is probably the most important factor contributing to the observed variability in annual fisheries landings (Challier et al., 2005a). The short lifespan of cephalopods means that stock size and commercial catches are heavily dependent on the number of juveniles successfully recruited to the population each year (Challier et al., 2002). However, a large degree of variability exists both in the timing of recruitment and in its success, with annual estimates ranging from 31 to 74 million recruits between 1995 and 2002 (Royer, 2002). An understanding of the factors affecting recruitment, and by inference population dynamics, is essential to the successful future management of this fishery. Recruitment variability is a complex phenomenon, related to both biotic and environmental factors (Gonzalez et al., 2010). As highlighted in this review, ELS are considered to be extremely responsive to changes in the ambient environment which are likely to affect their growth and/or survival (Boyle and Boletzky, 1996). With these, ELS considered significantly more vulnerable to mortality than their adult counterparts (Caddy, 1996). Temporal and/or spatial variation in growth and survival during ELS (particularly as a result of the effects of temperature) could explain some of the variability in the recruitment rate to the fishery for *S. officinalis* in the English Channel, as previously found in the fishery for *L. forbesi* in the same area (Challier et al., 2005b). For example, a link between mild winter conditions and cohort success, potentially regulated by differential ELS survival and growth, has already been demonstrated for both *L. forbesi* and *L. vulgaris* (Robin and Denis, 1999), suggesting that even small changes in the growth and/or survival rates of ELS can potentially have a large impact on subsequent recruitment rates (Pecl, 2004). Changes in juvenile growth rates and age-at-recruitment are likely to provide key parameters relative to recruitment strength (Challier et al., 2002).

4.2. Fecundity

In cephalopods, the accuracy of fecundity assessments is often affected by a lack of information concerning the exact spawning strategy of a species or population (e.g. whether they spawn over a single phase or over a series of intermittent phases; Mangold, 1987). If *S. officinalis* reproduces only once before dying, the effective individual fecundity of this species should be indicated by the number of eggs laid during the spawning event (Mangold, 1987). However, if intermittent spawning is occurring, then fecundity may be better reflected by an individual's potential fecundity which is measured from the total oocyte stock. In cephalopods, female fecundity has traditionally been calculated by measuring only the number of fully matured ova within an ovary, and not the number of eggs that are physically laid by an individual (Voss, 1983). Using this technique, the fecundity estimate of *S. officinalis* from mature ova range from 150 to 600 eggs (Mangold-Wirz, 1963; Richard, 1971). A wide range is observed for fecundity within a population and is accounted for at least in part by the size of an individual, such that the mean total number of ripe and advanced maturing eggs in the ovary increases from 99 to 543 eggs in animals of 80–190-mm ML

(Ezzedine-Najai, 1985). By counting the actual numbers of ova laid and then adding the number of ova that remained in the ovary following mortality, Boletzky (1975, 1987a) produced an estimate of fecundity for *S. officinalis* an order of magnitude higher than the aforementioned estimates at around 650 ova for small-sized females and 1000–2000 ova for large-sized females (~200-mm ML). This estimate has been further validated in more recent studies with several authors estimating the fecundity of captive female cuttlefish at around 3000 eggs (Forsythe et al., 1994; Hanley et al., 1998). While this estimate of fecundity seems high, individual fecundity can fluctuate considerably and may vary as a function of adult survival, dependent on whether the individual survives long enough during spawning to allow all available ova to complete maturation (Boletzky, 1987b). Thus, although the average number of fully mature ova is considered to be only 500 in a large female, aquarium observations such as Boletzky (1987b) suggests that if conditions permit for female survival, this fecundity can significantly increase as immature ova are given the time to mature ready for laying. Fecundity may therefore depend less on the functional capacity of the reproductive cycle than on other factors shortening the lifespan of females.

Fecundity not only can affect an individual's fitness but also can impact recruitment variability at the population level. Fecundity depends not only on the size of the female and the number of ova produced but also on the length of the spawning period and the length of survival of the individual following the onset of spawning (Boletzky, 1987b; Forsythe et al., 1994; Goff and Daguzan, 1991; Hanley et al., 1998).

4.3. Persistence of maternal effects

It has been suggested that the persistence of maternal effects can continue late in ontogeny and even into subsequent generations (Bernardo, 1996). Maternal effects can provide a mechanism for trans-generational phenotypic plasticity, in which the environment experienced by the mother can be translated into phenotypic variation in both her progeny and grandprogeny (e.g. Fox and Mousseau, 1998). For example, female oviposition decisions can affect not only the growth and development of her progeny but also that of her grandprogeny as these decisions can affect the environmental conditions that ELS encounter which have the capacity to affect their body size, condition and fecundity in later life, thus, in turn, effecting propagule size and resource allocation in the subsequent generation (Mousseau and Fox, 1998). However, maternal effects of oviposition site selection also have the potential to be more transitory in nature if, for example, paralarval dispersal were to occur, as ELS would move into different habitats or locations, effectively diluting the effects of maternal site selection. Thus, while the spatial distribution of natal sites can have a lasting impact on the distribution of older individuals and hence population dynamics at a large scale, whether distribution patterns are established at birth through maternal oviposition site selection (lay and stay) or are transient in nature (lay and leave) as a result of ELS dispersal behaviour will significantly affect the severity of this impact.

4.4. Age-at-recruitment

During the spring and summer months, embryonic and ELS cuttlefish are dispersed throughout the shallow coastal waters of the English Channel. During this time, ELS develop and grow and it is likely that individuals at different geographic or temporal locations will experience varying and dynamic environmental conditions during their time in these inshore waters. It is generally accepted that the main peak of recruitment to the English Channel cuttlefish fishery takes place in autumn (Challier et al., 2002). However, a study by Challier et al. (2005a) investigated differences in juvenile growth and age-at-recruitment in ELS cuttlefish from the English Channel. The authors sampled recruits on a monthly basis from October 2000 to June 2003 and in the summers of 2000 and 2002, pre-recruits from coastal waters (United Kingdom and France) were also sampled and the age of each individual determined through statolith analysis (Challier et al., 2005a). A significant difference was found in the growth rates of ELS cuttlefish from different geographical locations within these inshore waters. In 2002, the authors found that age of recruitment in the sampled cohort of English Channel *S. officinalis* varied between 2 and 5.5 months, with the majority ranging between 60 and 120 days of age (Challier et al., 2005a). Because the trawl fisheries, which target non-spawning cuttlefish both inshore and offshore, are size-selective (due to minimum mesh sizes), recruitment to the fishery is length dependent (Challier et al., 2005a). As a result, the differences in growth rates experienced by ELS at different spatial or temporal locations can influence their age-at-recruitment, with individuals in spawning grounds with conditions promoting faster growth rates, recruiting earlier than those in spawning grounds with conditions promoting a slower rate (Challier et al., 2005a) as already demonstrated for *Loligo gahi* in the Falkland Islands (Agnew et al., 2000, 2002). Thus, different spawning areas may be contributing differentially to recruitment, either at a given point in time or overall; if this is the case, then some spawning areas maybe acting as nursery grounds for this species, contributing a disproportionately high number of recruits to the population relative to other spawning areas. However, growth rates may also be influenced by other factors such as density dependence with the density of ELS cuttlefish in an area impacting on this parameter, through some form of competition for food or space (Challier et al., 2005a). The exact relationship between growth rate, age and recruitment therefore remains hard to determine and a continued development of time-series estimates of growth, age and recruitment of individuals from the natural population will be essential to develop our understanding of age-at-recruitment and the dominant factors affecting it (Challier et al., 2005a).

4.5. Timing of recruitment (onshore/offshore spawning)

Challier et al. (2005a) found that although the main period of recruitment occurred in autumn, a small amount of recruitment was also found to occur throughout the year as well. While age-at-recruitment was found to vary significantly between cohorts and seasons, it was consistently found to range between 3 and 4 months. While some hatching took place throughout the year, the majority hatched during the summer months (July and August) and subsequently recruited to the fishery in autumn (Challier et al., 2005a). However, although the majority of recruitment was shown to occur during the autumn, there were only small seasonal differences in the age-at-recruitment (e.g. 11–17 days) and by back-calculating the date of hatching, through statolith analysis, it was found that individuals were entering the fishery throughout the year and not just in autumn (Challier et al., 2005a). It had previously been assumed that all cuttlefish hatched during the summer months so that individuals recruiting to the fishery after the autumnal peak would be older than individuals which recruited during the autumn peak (Boucaud-Camou and Boismery, 1991). Challier et al. (2005a) found that for the sampled 2000 cohort, the opposite trend existed, and recruits continued to range between 3 and 4 months of age throughout the year. This pattern is known to occur in populations in the Mediterranean sea, where hatching occurs throughout the year with autumnal and summer peaks (Guerra and Castro, 1988). In order to explain these findings, the authors suggest two hypotheses:

1. Spawning could occur all year round, but with coastal spawning only occurring in the spring (Boucaud-Camou and Boismery, 1991; Wang et al., 2003).

2. Spawning could take place in both the spring and summer, but with adults laying both inshore (in warm, shallow waters) and offshore (in deeper, colder waters), as a result of temperature-dependent embryonic development and ELS growth, eggs laid coastally would then develop much quicker than those laid offshore, resulting in winter and spring hatching (Challier et al., 2005a).

A study by Bouchaud and Daguzan (1990) indicated that at 12 °C eggs of *S. officinalis* would take approximately 5 months to develop and that if the water temperature dropped below 9 °C, which is known to occur in this area during winter (Dunn, 1999), embryonic development would temporarily stop, restarting once the temperatures had increased. It is possible that spawning takes place in the English Channel in deeper waters and/or throughout the year. If offshore spawning does occur in this species, determining whether it occurs annually, or in response to a disturbance in recruitment rates from inshore spawning and whether it represents a small fraction of the reproductive output of the population or if it may have a significant effect on annual recruitment, is essential to aiding our understanding or population dynamics and variability in fisheries landings.

4.6. Juvenile and nursery habitats

Many commercially exploited cephalopods, including *S. officinalis*, live for less than 2 years, which means that recruitment success in these short-lived species is likely to be heavily dependent on both the physical and biological environments within their spawning and nursery grounds. Changes in survival rates of embryonic and ELS due to predation, starvation or inadequate physical conditions within a juvenile habitat are also potential sources of fluctuations in cephalopod recruitment (e.g. *L. forbesi*; Challier et al., 2006).

For many animals, an association with a particular habitat often leads to the assumption that they are exhibiting a 'preference' for it, with individuals actively 'selecting' it above all other available habitat types either at the phase of spawning or at hatching. The possibility exists that sexually mature female cuttlefish are actively assessing and choosing an optimal spawning with hatchlings remaining in their natal habitat following hatching (lay and stay hypothesis). Alternatively, spawning females may be distributed patchily along the English Channel coastline, spawning indiscriminately on the first structural habitat encountered and subsequently ELS redistributing to suitable micro-habitats within their natal habitat or to others further along the coastline (lay and leave hypothesis).

The suggestion that heterogeneous conditions within pre-recruit (e.g. spawning and juvenile) habitats may affect recruitment success, resulting in different habitats contributing in different proportions to the overall recruitment success of the population, is an idea addressed by Beck et al. (2001) in their illustration of the 'nursery-role hypothesis'. This hypothesis outlines the idea that certain areas of juvenile habitat may confer a selective advantage to the individuals within it, such as through optimal biotic and abiotic conditions, resulting in these areas contributing disproportionately to the population in terms of recruits (Beck et al., 2001). Such areas are known as nursery grounds and have been identified and described for many commercial marine species. While this has not yet been done for cuttlefish, the suggestion had been made in general terms that areas such as seagrass beds may act generically as nursery grounds for a wide number of commercial species, which may include coastal spawners like cuttlefish (e.g. Jackson et al., 2001).

Seagrass beds are highly productive nursery areas, with the potential to provide diverse and abundant sources of prey items and the structural complexity to provide shelter from predation to juveniles of many species (e.g. Jackson et al., 2001). Seagrasses are also recognised for their capacity to modify currents (e.g. Fonseca et al., 1982) and promote sediment deposition (e.g. Ginsburg and Lowenstam, 1958) providing a sandy substrate, which is useful for ELS to bury themselves, low wave exposure, which may reduce egg loss during embryonic development, depth or local warming, which may decrease embryonic development time. If key nursery areas can be identified for juvenile cuttlefish, then this may help direct future conservation and management strategies for this species.

4.7. Connectivity

The degree of connectivity between juvenile and adult habitats may also affect the rate of recruitment from a spawning ground and the success of individuals to recruit. In late autumn, ELS cuttlefish migrate from inshore coastal waters to the deeper waters in the central English Channel where they spend the colder winter months (Wang et al., 2003). It is possible that the distance or pathways from different locations or habitats can affect the number of juveniles recruiting to the stock, for example, the further a juvenile habitat is from the offshore wintering grounds, the longer the distance that an individual must travel and the greater the potential for predation or mortality enroute. Also, it may be that individuals from certain habitats where conditions are optimal for ELS growth and survival are bigger and better equipped for the migration than those from habitats with sub-optimal conditions for ELS growth and survival. However, many of these individuals will be of a recruitable size to the fishery during this offshore migration and so have the potential to be removed from the system via fisheries mortality as well as from natural mortality. Research has yet to be undertaken examining the movements and connectivity between inshore juvenile (or nursery) habitats and offshore wintering grounds for *S. officinalis* in the English Channel.

4.8. Climate change

The possible effects of climate change on the world's oceans include reduced salinity, increased temperatures (global warming) and increased acidity (ocean acidification) (e.g. Pörtner, 2008). The potential effects of climate change on marine organisms and ecosystems remain poorly understood (Vézina and Hoegh-Guldberg, 2008). It is accepted that global ocean pH has already fallen by 0.1 units and that it is likely to fall a further 0.3 units by 2050 (Feely et al., 2009). Ocean acidification can cause associated changes in the seawater carbonate chemistry leading to a reduction on calcium carbonate saturation in seawater (Zeebe and Wolf-Gladrow, 2001). This is of particular concern for calcifying marine organisms as this reduction could have a negative effect on their calcification process and metabolism (Gutowska et al., 2008; Pörtner, 2008).

The majority of invertebrates (e.g. cnidarians, molluscs and echinoderms) with low metabolic and activity rates responded negatively to elevated CO_2 concentrations (Fabry et al., 2008). However, in contrast to all previous invertebrate studies, Gutowska et al. (2008) showed that, as an active mollusc with a high metabolic rate, juvenile *S. officinalis* can maintain growth capacity and calcification of their internal aragonite shell over a 6-week period under elevated CO_2 conditions ($\sim$4000 and 6000 ppm). Test individuals were found to gain approximately 4% BW d^{-1}, maintain a metabolic rate of 0.09 μmol O_2 g^{-1} min^{-1} and increase their calcified cuttlebone mass by over 500% (Gutowska et al., 2008). These measurements were similar to those recorded for control subjects, although significantly higher calcification rates were recorded in individuals exposed to $\sim$6000 ppm CO_2 compared to the control group (Gutowska et al., 2008). The authors conclude that active cephalopods like *S. officinalis* must have a certain level of pre-adaptation to long-term increases in CO_2 levels enabling them to maintain (or increase) growth, metabolism and calcification rates (Gutowska et al., 2008).

The effect of CO_2 accumulation in the global ocean is not occurring in isolation and its interaction with effects of warming, eutrophication and hypoxia (oxygen deficiency arising from global warming and eutrophication) are currently of international concern (e.g. Pörtner et al., 2005). *S. officinalis* uses the blood pigment haemocyanin which has a limited capacity for carrying oxygen (3 mM) and thus rely on fully oxygenating haemocyanin at the gills and releasing the majority of the bound oxygen as it passes through the body (Melzner et al., 2007b). In *S. officinalis* at ambient temperatures of 17 °C, and under resting conditions, approximately 80% of bound oxygen is released as it passes through the body (Johansen et al., 1982). However, the oxygen binding properties of haemocyanin are highly temperature dependent and optimal affinity and oxygen transfer functions can only be maintained within a given thermal window (11–23 °C; Melzner et al., 2007b). Outside of this thermal window, saturation (in the warm) and desaturation (in the cold) of haemocyanin and its functioning is severely impaired and can lead to hypoxemia (progressive internal hypoxia) and death (Melzner et al., 2007b). This is known as the oxygen limitation of thermal tolerance and has been investigated in detail for *S. officinalis* (Melzner et al., 2006a,b, 2007a) and may pose a problem for this species in relation to the combined effects of climate change (increased water temperature and decreased oxygen saturation). Interestingly, the results of studies using *S. officinalis* have shown that smaller individuals are more hypoxia tolerant under thermal stress than larger ones, indicating a wider thermal window for smaller individuals (De Wachter et al., 1988; Johansen et al., 1982; Melzner et al., 2007a).

An understanding of the impact that climate change may have on the physiological processes of *S. officinalis* and on the marine ecosystems which they inhabit will be essential to model recruitment variability of this species and to enable realistic sustainable management to be developed.

5. SUMMARY

Variability in reproductive dynamics of *S. officinalis* is initially generated by the adults through variable egg production and quality and the timing and location of spawning events; these effects are then amplified or dampened as a result of the environmental conditions encountered by embryonic and paralarval phase in the spawning grounds which result from the preceding impact of maternal effects such as oviposition site selection.

The common cuttlefish *S. officinalis* is a short-lived, fast-growing species which is vulnerable to changes in environmental conditions, particularly during the ELS, which can significantly impact on annual recruitment strength. Annual recruitment success is essential and depends on breeding success and the survival of eggs to a recruitable age. It is thought to be largely driven by the suitability of environmental conditions for growth, survival and feeding within the ELS habitats. With only two overlapping generations, landings of *S. officinalis* from within the English Channel are also prone to large and unpredictable fluctuations as a result of the variable strength of annual recruitment. This makes it difficult for fisheries managers to determine the best course of action in order to ensure a viable and sustainable fishery for the future. As the English Channel cuttlefish fishery continues to expand, it is critical that we develop our understanding of the factors affecting ELS survival and recruitment in commercially exploited population. Such knowledge will help understand how annual recruitment strength can be predicted and recruitment failure avoided (i.e. reduced fishing during years of poor recruitment).

S. officinalis is known to exhibit high intra-specific plasticity in growth patterns and other life-history traits. Environmental conditions (which can vary both spatially and temporally) are thought to account for a significant portion of this variability, in particular, during the ELS of this species, water temperature, food availability, and size at hatching are considered to be the three key factors regulating growth rate and age-at-recruitment in the English Channel population of *S. officinalis*. The heterogeneity of these three factors is evident both spatially and temporally within different spawning grounds and habitats, but the effects of size at hatching are also impacted by parental genetics and maternal effects (e.g. maternal nutritional history).

Variability in the rates of ELS mortality (which has yet to be determined in natural populations of *S. officinalis*) will be critical in generating recruitment success or failure, and may vary significantly between locations and habitats. The potential for a particular habitat or location to be contributing disproportionately to the overall recruitment of this population does exist and highlights the potential for discrete nursery areas to occur (e.g. seagrass beds). Newly hatched individuals potentially experience significantly different predation and feeding pressures and/or environmental factors (e.g. temperature, salinity, etc.) depending on the conditions encountered at each spawning site. Another factor affecting embryonic mortality rates within this population is the loss of eggs which spawning females attach to fishing traps and which are removed from the traps during the middle or end of the season to reduce their weight and/or drag during hauling. An estimation of natural mortality rates, in particular, for those of ELS will be essential to understanding variations in annual recruitment rates for this species.

While *S. officinalis* is one of the better studied cephalopod species, we are still a long way from fully understanding the behaviour and ecology of this species in the wild, and, in particular, direct observations of ELS (pre-recruit stages) are limited. To date, the majority of information which provides the basis of our understanding of temperature-regulated growth in cephalopods has been determined by laboratory-based studies. Such studies are often undertaken under set (fixed) temperature regimes and without the interactions of other environmental factors. While they have provided valuable information on the effects of temperature on early life history parameters, they may not accurately reflect the situation in natural populations when temperature regimes are dynamic (seasonal changes in water temperature) and complicated or compounded by multiple additional environmental factors. Understanding how growth rates of ELS *S. officinalis* are determined is critical to research on stock assessments and sustainable exploitation of the fishery. Another limitation of the current literature is the gaps in our knowledge of natural populations. Marine species are by their nature difficult to observe in the wild; this situation is compounded in cephalopods as a result of their cryptic nature and the size-selective nature of the traditional fishery métiers (e.g. trawling and potting) or as a result of their presence in protected coastal habitats (e.g. seagrass beds), and is evident in the lack of natural observations of pre-recruit stages. Our knowledge of natural mortality, growth and feeding rates of ELS *S. officinalis* is very limited and we must rely on structured laboratory studies to continue to provide insights into the effects in natural populations. These gaps in our knowledge provide a major barrier to our understanding of the mechanisms driving recruitment and potential recruitment variability together with future effective fisheries management.

While the effects of environmental heterogeneity on ELS is a critical factor in determining annual recruitment success, to focus only on the impacts on embryonic and ELS would be at the detriment of the early life history as a whole. It is evident that variability in recruitment, through pre-recruit survival rates, is initially generated through the reproductive dynamics of the adults, through the timing and location of spawning, oviposition site selection and variable egg production or quality. It is therefore important that research is also conducted to investigate the movements and behaviour patterns of spawning adults, the factors responsible for migration patterns (e.g. natal homing) and the potential mechanisms involved in oviposition site selection. Novel electronic tagging methodologies are currently being tested for *S. officinalis* and may prove to provide the best means of obtaining such data (Bloor, 2012).

That conditions within ELS habitats can significantly affect the rate at which ELS recruit to the fishery is now well established and that heterogeneity in these conditions can occur both spatially (within the extent of the English Channel) and/or temporally (either throughout the season or between years), and that this is heavily impacted by the behaviours of spawning adults is also clear. Further studies are now required to identify key spawning grounds for this species within the English Channel and to ascertain whether nursery habitats exist. Such information will be crucial to the successful and sustainable management of this species both within the English Channel fishery and further afield.

ACKNOWLEDGMENTS

Funding support for this work was provided by the Interreg IV programme CRESH (Cephalopod Recruitment from English Channel Spawning Habitats), including a Ph.D studentship to Isobel Bloor.

REFERENCES

Agnew, D.J., Hill, S., Beddington, J.R., 2000. Predicting the recruitment strength of an annual squid stock: *Loligo gahi* around the Falkland Islands. Can. J. Fish. Aquat. Sci. 57, 2479–2487.

Agnew, D.J., Beddington, J.R., Hill, S.L., 2002. The potential use of environmental information to manage squid stocks. Can. J. Fish. Aquat. Sci. 59, 1851–1857.

Alves, C., Chichery, R., Boal, J.G., Dickel, L., 2006. Orientation in the cuttlefish *Sepia officinalis*: response versus place learning. Anim. Cogn. 10, 29–36.

Ambrose, R.F., 1988. Population dynamics of *Octopus bimaculatus*: influence of life history patterns, synchronous reproduction and recruitment. Malacologia 29, 23–39.

Arkhipkin, A., Laptikhovsky, V., 1994. Seasonal and interannual variability in growth and maturation of winter-spawning *Illex argentinus* (Cephalopoda, Ommastrephidae) in the Southwest Atlantic. Aquat. Living Resour. 7, 221–232.

Bachmann, M.D., Carlton, R.G., Burkholder, J.A.M., Wetzel, R.G., 1986. Symbiosis between salamander eggs and green algae: microelectrode measurements inside eggs demonstrate effect of photosynthesis on oxygen concentration. Can. J. Zool. 64, 1586–1588.

Baeza-Rojano, E., Garcìa, S., Garrido, D., Guerra-Garcìa, J.M., Domingues, P., 2010. Use of Amphipods as alternative prey to culture cuttlefish (*Sepia officinalis*) hatchlings. Aquaculture 300, 243–246.

Bateman, A.J., 1948. Intra-sexual selection in Drosophila. Heredity 2, 349–368.

Beck, M.W., Heck Jr., K.L., Able, K.W., Childers, D.L., Eggleston, D.B., Gillanders, B.M., Halpern, B., Hays, C.G., Hoshino, K., Minello, T.J., Orth, R.J., Sheridan, P.F., Weinstein, M.P., 2001. The identification, conservation, and management of estuarine and marine nurseries for fish and invertebrates. Bioscience 51, 633–641.

Begg, G.A., Friedland, K.A., Pearce, J.B., 1999. Stock identification and its role in stock assessment and fisheries management: an overview. Fish. Res. 43, 1–8.

Bernardo, J., 1996. Maternal effects in animal ecology. Am. Zool. 36, 83–105.

Blanc, A., 1998. Recehereches Bio-Ecologique Et Ecophysiologique De La Phase Juvenile De La Seiche *Sepia officinalis* Linne (Mollusque, Cephalopode, Sepiidae) Dans Le Golfe Du Morbihan (Sud Bretagne). Vie-Sante.

Blanc, A., Daguzan, J., 1998. Artificial surfaces for cuttlefish eggs (*Sepia officinalis* L.) in Morbihan Bay, France. Fish. Res. 38, 225–231.

Blanc, A., du Sel, G.P., Daguzan, J., 1998. Habitat and diet of early stages of *Sepia officinalis* L. (Cephalopoda) in Morbihan Bay, France. J. Molluscan. Stud. 64, 263–274.

Bloor, I., 2012. The Ecology, Distribution and Spawning Behaviour of the Commercially Important Common Cuttlefish (*Sepia officinalis*) in the Inshore Waters of the English Channel. Marine Institute and Marine Biological Association of the United Kingdom, Plymouth University.

Boal, J.G., 1997. Female choice of males in cuttlefish (Mollusca: Cephalopoda). Behaviour 134, 975–988.

Boal, J.G., Golden, D.K., 1999. Distance chemoreception in the common cuttlefish, *Sepia officinalis* (Mollusca, Cephalopoda). J. Exp. Mar. Biol. Ecol. 235, 307–317.

Boavida-Portugal, J., Moreno, A., Gordo, L., Pereira, J., 2010. Environmentally adjusted reproductive strategies in females of the commercially exploited common squid *Loligo vulgaris*. Fish. Res. 106, 193–198.

Boletzky, S.V., 1975. A contribution to the study of yolk absorption in the Cephalopoda. Zoomorphology 80, 229–246.

Boletzky, S.V., 1983. Sepia officinalis. In: Boyle, P.R. (Ed.), Cephalopod Life Cycles. Academic Press, London.

Boletzky, S.V., 1986a. Encapsulation of cephalopod embryos: a search for functional correlations. Am. Malacol. Bull. 4, 217–227.

Boletzky, S.V., 1986b. Reproductive strategies in cephalopods: variation and flexibility of life-history patterns. Adv. Invert. Reprod. 4, 379–389.

Boletzky, S.V., 1987a. Embryonic phase. In: Boyle, P.R. (Ed.), Cephalopod Life Cycles. Academic Press, London.

Boletzky, S.V., 1987b. Fecundity variation in relation to intermittent or chronic spawning in the cuttlefish, *Sepia officinalis* L. (Mollusca, Cephalopoda). Bull. Mar. Sci. 40, 382–387.

Boletzky, S.V., 1987c. Juvenile behaviour. In: Boyle, P.R. (Ed.), Cephalopod Life Cycles. Academic Press, London.

Boletzky, S.V., 1988. A new record of long-continued spawning in *Sepia officinalis* (Mollusca, Cephalopoda). Rapp. Comm. Int. Mer. Medit. 31, 257.

Boletzky, S.V., 1989. Recent studies on spawning, embryonic development, and hatching in the Cephalopoda. Adv. Mar. Biol. 25, 85–115.

Boletzky, S.V., 1994. Embryonic development of cephalopods at low temperatures. Antarctic Sci. 6, 139–142.

Boletzky, S.V., Erlwein, B., Hofmann, D.K., 2006. The Sepia egg: a showcase of cephalopod embryology. Vie Milieu 56, 191–201.

Boucaud-Camou, E., Boismery, J., 1991. The migrations of the cuttlefish (*Sepia officinalis* L.) in the English Channel. La Seiche: actes du Premier Symposium international sur la seiche. Caen, 1–3 Juin 1989. Centre de publications de l'universite de Caen.

Boucaud-Camou, E., Koueta, N., Boismery, J., Medhioub, A., 1991. The sexual cycle of *Sepia officinalis* L. from the Bay of Seine. La Seiche: actes du Premier Symposium international sur la seiche. Caen, 1–3 Juin 1989. Centre de publications de l'universite de Caen.

Bouchaud, O., 1991a. Energy consumption of the cuttlefish *Sepia officinalis* L. (Mollusca: Cephalopoda) during embryonic development, preliminary results. Bull. Mar. Sci. 49, 333–340.

Bouchaud, O., 1991b. Recherches physiologiques sur la reproduction de la seiche, *Sepia officinalis* (Mollusque, Cephalopode, Sepiidae), dans le secteur Mor Braz-Golfe du Morbihan (Sud Bretagne). l'université de Rennes.

Bouchaud, O., Daguzan, J., 1989. Etude du developpement de oeuf de *Sepia officinalis* L. (Cephalopode, Sepioidea) en conditions experimentales. Haliotis 19, 189–200.

Bouchaud, O., Daguzan, J., 1990. Etude experimentale de l'influence de la temperature sur le deroulement embryonnaire de la seiche *Sepia officinalis* L. (Cephalopoda, Sepioidae). Experimental study of temperature effects on the embryonic development of the cuttlefish *Sepia officinalis* L.(Cephalopoda, Sepioidae). Cah. Biol. Mar. 31, 131–145.

Bouchaud, O., Galois, R., 1990. Utilization of egg-yolk lipids during the embryonic development of *Sepia officinalis* L. in relation to temperature of the water. Comp. Biochem. Physiol. B 97, 611–615.

Boucher-Rodoni, R., Boucaud-Camou, E., Mangold, K., 1987. Feeding and digestion. In: Boyle, P.R. (Ed.), Cephalopod Life Cycles. Academic Press, London.

Boyle, P.R., 1990. Cephalopod biology in the fisheries context. Fish. Res. 8, 303–321.

Boyle, P.R., Boletzky, S.V., 1996. Cephalopod populations: definition and dynamics. Philos. Trans. Biol. Sci. 351, 985–1002.

Boyle, P.R., Rodhouse, P., 2005. Cephalopods: Ecology and Fisheries. Blackwell Science Ltd, Oxford.

Budelmann, B.U., 1994. Cephalopod sense organs, nerves and the brain: adaptations for high performance and life style. Mar. Behav. Physiol. 25, 13–33.

Budelmann, B.U., 1996. Active marine predators: the sensory world of cephalopods. Mar. Freshw. Behav. Physiol. 27, 59–75.

Bustamante, P., Teyssiè, J.L., Fowler, S., Cotret, O., Danis, B., Warnau, M., 2002. Biokinetics of cadmium and zinc accumulation and depuration at different stages in the life cycle of the cuttlefish *Sepia officinalis*. Mar. Ecol. Prog. Ser. 231, 167–177.

Bustamante, P., Teyssié, J.L., Danis, B., Fowler, S.W., Miramand, P., Cotret, O., Warnau, M., 2004. Uptake, transfer and distribution of silver and cobalt in tissues of the common cuttlefish *Sepia officinalis* at different stages of its life cycle. Mar. Ecol. Prog. Ser. 269, 185–195.

Bustamante, P., Teyssié, J.L., Fowler, S., Warnau, M., 2006. Contrasting bioaccumulation and transport behaviour of two artificial radionuclides (^{241}Am and ^{134}Cs) in cuttlefish eggshell. Vie Milieu 56, 153–156.

Caddy, J.F., 1983 The cephalopods: factors relevant to their population dynamics and to the assessment and management of stocks. FAO fisheries techical paper.

Caddy, J.F., 1996. Modelling natural mortality with age in short-lived invertebrate populations: definition of a strategy of genomonic time division. Aquat. Living Resour. 9, 197–207.

Calabrese, A., Nelson, D.A., 1974. Inhibition of embryonic development of the hard clam *Mercenaria mercenaria*, by heavy metals. Bull. Environ. Contam. Toxicol. 11, 92–97.

Caveriviére, A., Domain, F., Diallo, A., 1999. Observations on the influence of temperature on the length of embryonic development in *Octopus vulgaris* (Senegal). Aquat. Living Resour. 12, 151–154.

Centre for Environment Fisheries and Aquaculture Science, 2010. Sea temperature and salinity trends: presentation of results.

Challier, L., Royer, J., Robin, J.P., 2002. Variability in age-at-recruitment and early growth in English Channel *Sepia officinalis* described with statolith analysis. Aquat. Living Resour. 15, 303–311.

Challier, L., Dunn, M.R., Robin, J.P., 2005a. Trends in age-at-recruitment and juvenile growth of cuttlefish, *Sepia officinalis*, from the English Channel. ICES J. Mar. Sci. 62, 1671–1682.

Challier, L., Royer, J., Pierce, G.J., Bailey, N., Roel, B., Robin, J.P., 2005b. Environmental and stock effects on recruitment variability in the English Channel squid *Loligo forbesi*. Aquat. Living Resour. 18, 353–360.

Challier, L., Pierce, G.J., Robin, J.P., 2006. Spatial and temporal variation in age and growth in juvenile *Loligo forbesi* and relationships with recruitment in the English Channel and Scottish waters. J. Sea Res. 55, 217–229.

Choe, S., 1966. On the eggs, rearing, habits of the fry, and growth of some Cephalopoda. Bull. Mar. Sci. 16, 330–348.

Cohen, C.S., Strathmann, R.R., 1996. Embryos at the edge of tolerance: effects of environment and structure of egg masses on supply of oxygen to embryos. Biol. Bull. 190, 8–15.

Concise Oxford Dictionary, 1999. Concise Oxford Dictionary. Oxford University Press.

Correia, M., Palma, J., Andrade, J.P., 2008a. Effects of live prey availability on growth and survival in the early stages of cuttlefish Sepia officinalis (Linnaeus, 1758) life cycle. Aquacult. Res. 39, 33–40.

Correia, M., Palma, J., Kirakowski, T., Andrade, J.P., 2008b. Effects of prey nutritional quality on the growth and survival of juvenile cuttlefish, Sepia officinalis (Linnaeus, 1758). Aquacult. Res. 39, 869–876.

Cowan, J.H., Shaw, R.F., 2002. Recruitment. In: Fuiman, L.A., Werner, R.G. (Eds.), Fishery Science: The Unique Contribution of Early Life Stages. Blackwell Science Ltd, Oxford.

Cronin, E.R., Seymour, R.S., 2000. Respiration of the eggs of the giant cuttlefish Sepia apama. Mar. Biol. 136, 863–870.

Darmaillacq, A.S., Chichery, R., Poirier, R., Dickel, L., 2004. Effect of early feeding experience on subsequent prey preference by cuttlefish, Sepia officinalis. Dev. Psychobiol. 45, 239–244.

Darmaillacq, A.S., Chichery, R., Dickel, L., 2006a. Food imprinting, new evidence from the cuttlefish Sepia officinalis. Biol. Lett. 2, 345–347.

Darmaillacq, A.S., Chichery, R., Shashar, N., Dickel, L., 2006b. Early familiarization overrides innate prey preference in newly hatched Sepia officinalis cuttlefish. Anim. Behav. 71, 511–514.

Darmaillacq, A.S., Lesimple, C., Dickel, L., 2008. Embryonic visual learning in the cuttlefish, Sepia officinalis. Anim. Behav. 76, 131–134.

Denis, V., Robin, J.P., 2000. Spatio-temporal patterns in cephalopod resources exploited by the French Atlantic Fishery-Empirical models of abundance based on environmental parameters. ICES J. Mar. Sci. 59, 633–848.

Denis, V., Robin, J.P., 2001. Present status of the French Atlantic fishery for cuttlefish (Sepia officinalis). Fish. Res. 52, 11–22.

Denis, V., Lejeune, J., Robin, J.P., 2002. Spatio-temporal analysis of commercial trawler data using general additive models: patterns of Loliginid squid abundance in the north-east Atlantic. ICES J. Mar. Sci. 59, 633–648.

Derby, C.D., Kicklighter, C.E., Johnson, P.M., Zhang, X., 2007. Chemical composition of inks of diverse marine molluscs suggests convergent chemical defenses. J. Chem. Ecol. 33, 1105–1113.

Derusha, R.H., Forsythe, J.W., Dimarco, F.P., Hanlon, R.T., 1989. Alternative diets for maintaining and rearing cephalopods in captivity. Lab. Anim. Sci. 39, 306.

De Wachter, B., Wolf, G., Richard, A., Decleir, W., 1988. Regulation of respiration during juvenile development of Sepia officinalis (Mollusca: Cephalopoda). Mar. Biol. 97, 365–371.

Dickel, L., Chichery, M.P., Chichery, R., 1997. Postembryonic maturation of the vertical lobe complex and early development of predatory behavior in the cuttlefish (Sepia officinalis). Neurobiol. Learn. Mem. 67, 150–160.

Dingle, H., 1996. Migration: The Biology of Life on the Move. Oxford University Press, New York.

Dingle, H., Drake, V.A., 2007. What is migration? Bioscience 57, 113–121.

Dodson, J.J., 1997. Fish migration: an evolutionary perspective. In: Godin, J. (Ed.), Behavioural Ecology of Teleost Fish. Oxford University Press, Oxford.

Domingues, P.M., Kingston, T., Sykes, A., Andrade, J.P., 2001a. Growth of young cuttlefish, Sepia officinalis (Linnaeus 1758) at the upper end of the biological distribution temperature range. Aquacult. Res. 32, 923–930.

Domingues, P.M., Sykes, A., Andrade, J.P., 2001b. The use of Artemia sp. or mysids as food source for hatchlings of the cuttlefish (Sepia officinalis L.); effects on growth and survival throughout the life cycle. Aquacul. Int. 9, 319–331.

Domingues, P., Poirier, R., Dickel, L., Almansa, E., Sykes, A., Andrade, J.P., 2003. Effects of culture density and live prey on growth and survival of juvenile cuttlefish, Sepia officinalis. Aquacult. Int. 11, 225–242.

Domingues, P., Sykes, A., Sommerfield, A., Almansa, E., Lorenzo, A., Andrade, J.P., 2004. Growth and survival of cuttlefish (Sepia officinalis) of different ages fed crustaceans and fish. Effects of frozen and live prey. Aquaculture 229, 239–254.

Donohue, K., 1999. Seed dispersal as a maternally influenced character: mechanistic basis of maternal effects and selection on maternal characters in an annual plant. Am. Nat. 154, 674–689.

Dunn, M.R., 1999. Aspects of the stock dynamics and exploitation of cuttlefish, Sepia officinalis (Linnaeus, 1758), in the English Channel. Fish. Res. 40, 277–293.

Eberhard, W.G., 1985. Sexual Selection and Animal Genitalia. Harvard University Press, Cambridge.

Eberhard, W.G., 1996. Female control: sexual selection by cryptic female choice. Princeton University Press, Princeton.

Ezzedine-Najai, S., 1985. Fecundity of cuttlefish, Sepia officinalis L. (Mollusca: Cephalopoda) from the gulf of Tunis. Vie Milieu 35, 283–284.

Fabry, V.J., Seibel, B.A., Feely, R.A., Orr, J.C., 2008. Impacts of ocean acidification on marine fauna and ecosystem processes. ICES J. Mar. Sci. 65, 414–432.

Feely, R.A., Doney, S.C., Cooley, S.R., 2009. Ocean acidification: present conditions and future changes in a high-CO_2 world. Oceanography 22, 36–47.

Fonseca, M.S., Fisher, J.S., Zieman, J.C., Thayer, G.W., 1982. Influence of the seagrass, Zostera marina L., on current flow. Estuar. Coast. Shelf Sci. 15, 351–358.

Food and Agriculture Organisation of the United Nations (FAO), 2010. Yearbook. Fishery and Aquaculture Statistics. 2008. Statistics, of the Fisheries, Information Service Department, Aquaculture. Fishery and Aquaculture Statistics.

Food and Agriculture Organisation of the United Nations (FAO), 1964. Yearbook Fisheries Statistics. Catches and landings 1963. Statistics, of the Fisheries, Information Service Department, Aquaculture. Fishery and Aquaculture Statistics.

Food and Agriculture Organisation of the United Nations (FAO), 2012. Global Production Statistics 1950–2010.

Forsythe, J.W., 1993. A working hypothesis of how seasonal temperature change may impact the field growth of young cephalopods. In: Okutani, T., O'Dor, R.K., Kubodera, T. (Eds.), Recent Advances in Cephalopod Fisheries Biology. Tokai University Press, Tokyo, pp. 133–143.

Forsythe, J.W., 2004. Accounting for the effect of temperature on squid growth in nature: from hypothesis to practice. Mar. Freshw. Res. 55, 331–339.

Forsythe, J.W., Hanlon, R.T., 1988. Effect of temperature on laboratory growth, reproduction and life span of Octopus bimaculoides. Mar. Biol. 98, 369–379.

Forsythe, J.W., Hanlon, R.T., 1989. Growth of the eastern Atlantic squid, Loligo forbesi Steenstrup (Mollusca: Cephalopoda). Aquacult. Res. 20, 1–14.

Forsythe, J.W., Van Heukelem, W.F., 1987. Growth. In: Boyle, P.R. (Ed.), Cephalopod Life Cycles. Academic Press, London.

Forsythe, J.W., Derusha, R.H., Hanlon, R.T., 1994. Growth, reproduction and life span of Sepia officinalis (Cephalopoda: Mollusca) cultured through seven consecutive generations. J. Zool. Lond. 233, 175–192.

Forsythe, J.W., Walsh, L.S., Turk, P.E., Lee, P.G., 2001. Impact of temperature on juvenile growth and age at first egg-laying of the Pacific reef squid Sepioteuthis lessoniana reared in captivity. Mar. Biol. 138, 103–112.

Forsythe, J., Lee, P., Walsh, L., Clark, T., 2002. The effects of crowding on growth of the European cuttlefish, *Sepia officinalis* Linnaeus, 1758 reared at two temperatures. J. Exp. Mar. Biol. Ecol. 269, 173–185.

Fox, C.W., Mousseau, T.A., 1998. Maternal effects as adaptations for transgenerational phenotypic plasticity in insects. In: Mousseau, T.A., Fox, C.W. (Eds.), Maternal Effects as Adaptations, vol. 159. Oxford University Press, New York.

Franz, M.O., Mallot, H.A., 2000. Biomimetic robot navigation. Robot. Auton. Sys. 30, 133–153.

Fuiman, L.A., 2002. Special considerations of fish eggs and larvae. In: Fuiman, L.A., Werner, R.G. (Eds.), Fishery Science: The Unique Contributions of Early Life Stages. Blackwell Science Ltd, Oxford.

Gauvrit, E., Goff, R.L., Daguzan, J., 1997. Reproductive cycle of the cuttlefish *Sepia officinalis* (L) in the northern part of the Bay of Biscay. J. Molluscan. Stud. 63, 19–28.

Ginsburg, R.N., Lowenstam, H.A., 1958. The influence of marine bottom communities on the depositional environment of sediments. J. Geol. 66, 310–318.

Goff, R.L., Daguzan, J., 1991. Growth and life cycles of the cuttlefish *Sepia officinalis* L. (Mollusca: Cephalopoda) in South Brittany (France). Bull. Mar. Sci. 49, 341–348.

Gonzalez, A.F., Otero, J., Pierce, G.J., Guerra, A., 2010. Age, growth, and mortality of *Loligo vulgaris* wild paralarvae: implications for understanding of the life cycle and longevity. ICES J. Mar. Sci. 67, 1119–1127.

Grimpe, G., 1926. Biologische beobachtungen an *Sepia officinalis*. Verh. dt. zool. Ges. 31, 148–153 (Suppl. Zool. Anz.).

Guerra, A., 2006. Ecology of *Sepia officinalis*. Vie Milieu 56, 97–107.

Guerra, A., Castro, B.G., 1988. On the life of *Sepia officinalis* (Cephalopoda, Sepioidea) in the Ria de Vigo (NW Spain). Cah. Biol. Mar. 29, 395–405.

Guerra, A., Castro, B.G., 1994. Reproductive-somatic relationships in *Loligo gahi* (Cephalopoda: Loliginidae) from the Falkland Islands. Antarctic Sci. 6, 175–178.

Guerra, A., González, J.L., 2011. First record of predation by a tompot blenny on the common cuttlefish *Sepia officinalis* eggs. Vie Milieu 61, 45–48.

Guerra-Garcìa, J.M., Tierno de Figueroa, J.M., 2009. What do caprellids (Crustacea: Amphipoda) feed on? Mar. Biol. 156, 1881–1890.

Gutowska, M.A., Melzner, F., 2009. Abiotic conditions in cephalopod (*Sepia officinalis*) eggs: embryonic development at low pH and high pCO_2. Mar. Biol. 156, 515–519.

Gutowska, M.A., Pörtner, H.-O., Melzner, F., 2008. Growth and calcification in the cephalopod *Sepia officinalis* under elevated seawater pCO_2. Mar. Ecol. Prog. Ser. 373, 303–309.

Hall, K.C., Hanlon, R.T., 2002. Principal features of the mating system of a large spawning aggregation of the giant Australian cuttlefish *Sepia apama* (Mollusca: Cephalopoda). Mar. Biol. 140, 533–545.

Hanley, J.S., Shashar, N., Smolowitz, R., Bullis, R.A., Mebane, W.N., Gabr, H.R., Hanlon, R.T., 1998. Modified laboratory culture techniques for the European cuttlefish *Sepia officinalis*. Biol. Bull. 195, 223–225.

Hanlon, R.T., Messenger, J.B., 1988. Adaptive coloration in young cuttlefish (*Sepia officinalis* L.): the morphology and development of body patterns and their relation to behaviour. Philos. Trans. R. Soc. Lond. B Biol. Sci. 320, 437–487.

Hanlon, R.T., Messenger, J.B., 1996. Cephalopod Behaviour. Cambridge University Press, Cambridge.

Hanlon, R.T., Ament, S.A., Gabr, H., 1999. Behavioral aspects of sperm competition in cuttlefish, *Sepia officinalis* (Sepioidea: Cephalopoda). Mar. Biol. 134, 719–728.

Hastie, L.C., Pierce, G.J., Wang, J., Bruno, I., Moreno, A., Piatkowski, U., Robin, J.P., 2009. Cephalopods in the North-Eastern Atlantic: species, biogeography, ecology, exploitation and conservation. Oceanogr. Mar. Biol. 47, 111–190.

Hatfield, E., 2000. Do some like it hot? Temperature as a possible determinant of variability in the growth of the Patagonian squid, *Loligo gahi* (Cephalopoda: Loliginidae). Fish. Res. 47, 27–40.

Hatfield, E.M.C., Hanlon, R.T., Forsythe, J.W., Grist, E.P.M., 2001. Laboratory testing of a growth hypothesis for juvenile squid *Loligo pealeii* (Cephalopoda: Loliginidae). Can. J. Fish. Aquat. Sci. 58, 845–857.

Healy, S.D., 2006. Imprinting: seeing food and eating it. Curr. Biol. 16, R501–R502.

Ibáñez, C.M., Keyl, F., 2010. Cannibalism in cephalopods. Rev. Fish Biol. Fish. 20, 123–136.

Ikeda, Y., Itoo, K., Matsumotop, G., 2004. Does light intensity affect embryonic development of squid (*Heterololigo bleekeri*)? J. Mar. Biol. Assoc. U.K. 84, 1215–1219.

International Council for Exploration of the Sea (ICES), 2003. Report of the Working Group on Cephalopod Fisheries and Life History (WGCEPH), 4–6 December 2002, Lisbon, Portugal.

International Council for Exploration of the Sea (ICES), 2010. Report of the Working Group on Cephalopod Fisheries and Life History (WGCEPH), 9–11 March 2010, Sukarrieta, Spain.

Jackson, G.D., 2004. Cephalopod growth: historical context and future directions. Mar. Freshw. Res. 55, 327–329.

Jackson, G., Moltschaniwskyj, N., 2002. Spatial and temporal variation in growth rates and maturity in the Indo-Pacific squid *Sepioteuthis lessoniana* (Cephalopoda: Loliginidae). Mar. Biol. 140, 747–754.

Jackson, G.D., Forsythe, J.W., Hixon, R.F., Hanlon, R.T., 1997. Age, growth, and maturation of *Lolliguncula brevis* (Cephalopoda: Loliginidae) in the northwestern Gulf of Mexico with a comparison of length-frequency versus statolith age analysis. Can. J. Fish. Aquat. Sci. 54, 2907–2919.

Jackson, E.L., Rowden, A.A., Attrill, M.J., Bossey, S.J., Jones, M.B., 2001. The importance of seagrass beds as a habitat for fishery species. Oceanogr. Mar. Biol. 39, 269–304.

Jereb, P., Roper, C.F.E., 2005. Cephalopods of the World: An Annotated and Illustrated Catalogue of Cephalopod Species Known to Date. FAO Species Catalogue for Fishery Purposes No. 4, vol. 1. Food and Agriculture Organisation, Rome.

Johansen, K., Brix, O., Lykkeboe, G., 1982. Blood gas transport in the cephalopod, *Sepia officinalis*. J. Exp. Biol. 99, 331–338.

Karson, M.A., Boal, J.G., Hanlon, R.T., 2003. Experimental evidence for spatial learning in cuttlefish (*Sepia officinalis*). J. Comp. Psychol. 117, 149–155.

Kerrigan, B.A., 1997. Variability in larval development of the tropical reef fish *Pomacentrus amboinensis* (Pomacentridae): the parental legacy. Mar. Biol. 127, 395–402.

Koueta, N., Boucaud-Camou, E., 1999. Food intake and growth in reared early juvenile cuttlefish *Sepia officinalis* L. (Mollusca: Cephalopoda). J. Exp. Mar. Biol. Ecol. 240, 93–109.

Koueta, N., Boucaud-Camou, E., 2001. Basic growth relations in experimental rearing of early juvenile cuttlefish *Sepia officinalis* L. (Mollusca: Cephalopoda). J. Exp. Mar. Biol. Ecol. 265, 75–87.

Koueta, N., Boucaud-Camou, E., 2003. Combined effects of photoperiod and feeding frequency on survival and growth of juvenile cuttlefish *Sepia officinalis* L. in experimental rearing. J. Exp. Mar. Biol. Ecol. 296, 215–226.

Koueta, N., Camou-Boucaud, E., Renou, A.M., 1995. Gonadotropic mitogenic activity of the optic gland of the cuttlefish, *Sepia officinalis*, during sexual maturation. J. Mar. Biol. Assoc. U.K. 75, 391–404.

Koueta, N., Castro, B.G., Boucaud-Camou, E., 2000. Biochemical indices for instantaneous growth estimation in young cephalopod *Sepia officinalis* L. ICES J. Mar. Sci. 57, 1–7.

Koueta, N., Boucaud-Camou, E., Noel, B., 2002. Effect of enriched natural diet on survival and growth of juvenile cuttlefish *Sepia officinalis* L. Aquaculture 203, 293–310.

Lacoue-Labarthe, T., Warnau, M., Oberhansli, F., Teyssiè, J.L., Jeffree, R., Bustamante, P., 2008a. First experiments on the maternal transfer of metals in the cuttlefish *Sepia officinalis*. Mar. Pollut. Bull. 57, 826–831.

Lacoue-Labarthe, T., Warnau, M., Oberhansli, F., Teyssiè, J.L., Koueta, N., Bustamante, P., 2008b. Differential bioaccumulation behaviour of Ag and Cd during the early development of the cuttlefish *Sepia officinalis*. Aquat. Toxicol. 86, 437–446.

Lacoue-Labarthe, T., Martin, S., Oberhansli, F., Teyssié, J.L., Markich, S., Jeffree, R., Bustamante, P., 2009. Effects of increased pCO_2 and temperature on trace element (Ag, Cd and Zn) bioaccumulation in the eggs of the common cuttlefish, *Sepia officinalis*. Biogeosciences 6, 2561–2573.

Lacoue-Labarthe, T., Warnau, M.F., Oberhänsli, F., Teyssié, J.L., Bustamante, P., 2010. Contrasting accumulation biokinetics and distribution of [241]Am, Co, Cs, Mn and Zn during the whole development time of the eggs of the common cuttlefish, *Sepia officinalis*. J. Exp. Mar. Biol. Ecol. 382, 131–138.

Lamare, M.D., Barker, M.F., 1999. *In situ* estimates of larval development and mortality in the New Zealand sea urchin *Evechinus chloroticus* (Echinodermata: Echinoidea). Mar. Ecol. Prog. Ser. 180, 197–211.

Langridge, K.V., Broom, M., Osorio, D., 2007. Selective signalling by cuttlefish to predators. Curr. Biol. 17, R1044–R1045.

Laptikhovsky, V., Salman, A., Onsoy, B., Katagan, T., 2003. Fecundity of the common cuttlefish *Sepia officinalis* L. (Cephalopoda, Sepiida): a new look at the old problem. Sci. Mar. 67, 279–284.

Lee, P.G., 1994. Nutrition of cephalopods: fueling the system. Mar. Freshw. Behav. Physiol. 25, 35–51.

Leporati, S.C., Pecl, G.T., Semmens, J.M., 2007. Cephalopod hatchling growth: the effects of initial size and seasonal temperatures. Mar. Biol. 151, 1375–1383.

Lohmann, K.J., Putman, N.F., Lohmann, C.M.F., 2008. Geomagnetic imprinting: a unifying hypothesis of long-distance natal homing in salmon and sea turtles. Proc. Natl. Acad. Sci. 105, 19096–19101.

Mangold, K., 1987. Reproduction. In: Boyle, P.R. (Ed.), Cephalopod Life Cycles. Academic Press, London.

Mangold-Wirz, K., 1963. Biologie des Cephalopodes benthiques et nectoniques de la Mer Catalane. Vie Milieu 13, 1–285.

Marine Management Organisation (MMO), 2010. The UK Fishing Industry in 2010 Landings. Marine Management Organisation, London.

Marshall, D.J., Bolton, T.F., Keough, M.J., 2003. Offspring size affects the post-metamorphic performance of a colonial marine invertebrate. Ecology 84, 3131–3137.

Marshall, D.J., Allen, R.M., Crean, A.J., 2008. The ecological and evolutionary importance of maternal effects in the sea. Oceanogr. Mar. Biol. Annu. Rev. 46, 203–250.

McLay, C.L., Guinot, D., 1997. Ten arms meet ten legs: Decapoda (Mollusca: Cephalopoda: Sepioidea) Spawn on Decapoda (Crustacea: Brachyura: Homolidae). J. Crustacean Biol. 17, 692–694.

Meekan, M.G., Fortier, L., 1996. Selection for fast growth during the larval life of Atlantic cod *Gadus morhua* on the Scotian Shelf. Mar. Ecol. Prog. Ser. 137, 25–37.

Meekan, M.G., Vigliola, L., Hansen, A., Doherty, P.J., Halford, A., Carleton, J.H., 2006. Bigger is better: size-selective mortality throughout the life history of a fast-growing clupeid, *Spratelloides gracilis*. Mar. Ecol. Prog. Ser. 317–324, 237.

Melzner, F., Bock, C., Pörtner, H.-O., 2006a. Critical temperatures in the cephalopod *Sepia officinalis* investigated using *in vivo* ^{31}P NMR spectroscopy. J. Exp. Biol. 209, 891–906.

Melzner, F., Bock, C., Pörtner, H.-O., 2006b. Temperature-dependent oxygen extraction from the ventilatory current and the costs of ventilation in the cephalopod *Sepia officinalis*. J. Comp. Physiol. B 176, 607–621.

Melzner, F., Bock, C., Pörtner, H.-O., 2007a. Allometry of thermal limitation in the cephalopod *Sepia officinalis*. Comp. Biochem. Physiol. A 146, 149–154.

Melzner, F., Mark, F.C., Pörtner, H.-O., 2007b. Role of blood-oxygen transport in thermal
tolerance of the cuttlefish, *Sepia officinalis*. Integrative Comp. Biol. 47, 645–655.

Messenger, J.B., 2001. Cephalopod chromatophores: neurobiology and natural history. Biol.
Rev. 76, 473–528.

Miramand, P., Bustamante, P., Bentley, D., Kouèta, N., 2006. Variation of heavy metal concentrations (Ag, Cd, Co, Cu, Fe, Pb, V, and Zn) during the life cycle of the common
cuttlefish *Sepia officinalis*. Sci. Total Environ. 361, 132–143.

Moltschaniwskyj, N.A., Pecl, G.T., 2003. Small-scale spatial and temporal patterns of egg production by the temperate loliginid squid *Sepioteuthis australis*. Mar. Biol. 142, 509–516.

Moltschaniwskyj, N., Pecl, G., Lyle, J., Haddon, M., Steer, M., 2003. Population dynamics
and reproductiven ecology of the southern calamary (*Sepioteuthis australis*) in Tasmania.
FRDC final report.

Morte, S., Redon, M.J., Sanz-Brau, A., 1997. Feeding habits of juvenile *Mustelus mustelus*
(Carcharhiniformes, Triakidae) in the western Mediterranean. Cah. Biol. Mar. 38,
103–107.

Mousseau, T.A., Dingle, H., 1991. Maternal effects in insect life histories. Annu. Rev.
Entomol. 36, 511–534.

Mousseau, T.A., Fox, C.W., 1998. The adaptive significance of maternal effects. Trends
Ecol. Evol. 13, 403–407.

Naud, M.J., Shaw, P.W., Hanlon, R.T., Havenhand, J.N., 2005. Evidence for biased use of
sperm sources in wild female giant cuttlefish (*Sepia apama*). Proc. R. Soc. Lond. B Biol.
Sci. 272, 1047–1051.

Nixon, M., Mangold, K., 1998. The early life of *Sepia officinalis*, and the contrast with that of
Octopus vulgaris (Cephalopoda). J. Zool. 245, 407–421.

Odling-Smee, L., Braithwaite, V.A., 2003. The role of learning in fish orientation. Fish Fish.
4, 235–246.

Palmegiano, G.B., D'Apote, M.P., 1983. Combined effects of temperature and salinity on
cuttlefish (*Sepia officinalis* L.) hatching. Aquaculture 35, 259–264.

Pascual, E., 1978. Crecimiento y alimentacion de tres generaciones de *Sepia officinalis* en cul-
tivo. Invest. Pesq. 42, 421–441.

Paulij, W.P., Bogaards, R.H., Denuce, J.M., 1990a. Influence of salinity on embryonic
development and the distribution of *Sepia officinalis* in the Delta Area (South Western
part of The Netherlands). Mar. Biol. 107, 17–23.

Paulij, W.P., Herman, P.M.J., van Hannen, E.J., Denuce, J.M., 1990b. The impact of photoperiodicity on hatching of *Loligo vulgaris* and *Loligo forbesi*. J. Mar. Biol. Assoc. U.K. 70,
597–610.

Paulij, W.P., Herman, P.M.J., Roozen, M.E.F., Denuce, J.M., 1991. The influence
of photoperiodicity on hatching of *Sepia officinalis*. J. Mar. Biol. Assoc. U.K. 71, 665–678.

Pawson, M.G., 1995. Biogeographical identification of English Channel fish and shellfish
stocks. MAFF fisheries research technical report, 99.

Pechenik, J.A., Eyster, L.S., Widdows, J., Bayne, B.L., 1990. The influence of food concentration and temperature on growth and morphological differentiation of blue mussel
Mytilus edulis L. larvae. J. Exp. Mar. Biol. Ecol. 136, 47–64.

Pecl, G.T., 2001. Flexible reproductive strategies in tropical and temperate Sepioteuthis
squids. Mar. Biol. 138, 93–101.

Pecl, G.T., 2004. The *in situ* relationships between season of hatching, growth and condition
in the southern calamary, *Sepioteuthis australis*. Mar. Freshw. Res. 55, 429–438.

Pecl, G.T., Steer, M.A., Hodgson, K.E., 2004. The role of hatchling size in generating the
intrinsic size-at-age variability of cephalopods: extending the Forsythe Hypothesis. Mar.
Freshw. Res. 55, 387–394.

Pepin, P., 1988. Predation and starvation of larval fish: a numerical experiment of size-and growth-dependent survival. Biol. Oceanogr. 6, 23–44.

Perez-Losada, M., Guerra, A., Sanjuan, A., 1999. Allozyme differentiation in the cuttlefish *Sepia officinalis* (Mollusca: Cephalopoda) from the NE Atlantic and Mediterranean. Heredity 83, 280–289.

Perez-Losada, M., Guerra, A., Carvalho, G.R., Sanjuan, A., Shaw, P.W., 2002. Extensive population subdivision of the cuttlefish *Sepia officinalis* (Mollusca: Cephalopoda) around the Iberian Peninsula indicated by microsatellite DNA variation. Heredity 89, 417–424.

Piatkowski, U., Pierce, G.J., Morais DA Cunha, M., 2001. Impact of cephalopods in the food chain and their interaction with the environment and fisheries: an overview. Fish. Res. 52, 5–10.

Pierce, G.J., Guerra, A., 1994. Stock assessment methods used for cephalopod fisheries. Fish. Res. 21, 255–285.

Pierce, G.J., Valavanis, V.D., Guerra, A., Jereb, P., Orsi-Relini, L., Bellido, J.M., Katara, I., Piatkowski, U., Pereira, J., Balguerias, E., Sobrino, I., Lefkaditou, E., Wang, J., Santurtun, M., Boyle, P.R., Hastie, L.C., Macleod, C.D., Smith, J.M., Viana, M., González, Angel F., Zuur, A.F., 2008. A review of cephalopod environment interactions in European Seas. Hydrobiologia 612, 49–70.

Pierce, G.J., Allcock, L., Bruno, I., Bustamante, P., Gonzalez, A., Guerra, Â., Jereb, P., Lefkaditou, E., Malham, S., Moreno, A., 2010. Cephalopod biology and fisheries in Europe. ICES cooperative research report, 303.

Pinder, A., Friet, S., 1994. Oxygen transport in egg masses of the amphibians *Rana sylvatica* and *Ambystoma maculatum*: convection, diffusion and oxygen production by algae. J. Exp. Biol. 197, 17.

Poirier, R., Chichery, R., Dickel, L., 2004. Effects of rearing conditions on sand digging efficiency in juvenile cuttlefish. Behav. Processes 67, 273–279.

Poole, H.H., Atkins, W.R.G., 1937. The penetration into the sea of light of various wavelengths as measured by emission or rectifier photo-electric cells. Proc. R. Soc. Lond. B Biol. Sci. 123, 151–165.

Pörtner, H.-O., 2008. Ecosystem effects of ocean acidification in times of ocean warming: a physiologist's view. Mar. Ecol. Prog. Ser. 373, 203–217.

Pörtner, H.-O., Langenbuch, M., Michaelidis, B., 2005. Synergistic effects of temperature extremes, hypoxia, and increases in CO_2 on marine animals: from Earth history to global change. J. Geophys. Res. 110, .

Resetarits Jr., W.J., 1996. Oviposition site choice and life history evolution. Am. Zool. 36, 205–215.

Richard, A., 1966a. Action de la temperature sur l'evolution genitale de *Sepia officinalis* L. Comptes Rendus de l'Academie des Science de Paris 263, 1998–2001.

Richard, A., 1966b. La temperature, facteur externe essentiel de croissance pour le Cephalopode *Sepia officinalis* L. Compte rendu de là Académide des Sciences de Paris Série D 263, 1138–1141.

Richard, A., 1971. Contribution a l'etude de la biologie de la croissance et de la maturation sexuelle de *Sepia officinalis* L. (Mollusque Cephalopode). University of Lille.

Roberts, M.J., Sauer, W.H.H., 1994. Environment: the key to understanding the South African chokka squid (*Loligo vulgaris reynaudii*) life cycle and fishery? Antarctic Sci. 6, 249–258.

Robin, J.P., Denis, V., 1999. Squid stock fluctuations and water temperature: temporal analysis of English Channel Loliginidae. J. Appl. Ecol. 36, 101–110.

Rocha, F., Guerra, Â., Gonzalez, Â.F., 2001. A review of reproductive strategies in cephalopods. Biol. Rev. 76, 291–304.

Rodhouse, P.G., 2001. Managing and forecasting squid fisheries in variable environments. Fish. Res. 54, 3–8.

Roper, C.F.E., Sweeney, M.J., Nauen, C., 1984. Cephalopods of the World: An Annotated and Illustrated Catalogue of Species of Interest to Fisheries. FAO Fisheries Synopsis, No. 125, vol. 3. Food and Agriculture Organisation, Rome.

Royer, J., 2002. Modelisation des Stocks de Cephalopodes de Manche. University of Caen, France.

Royer, J., Pierce, G.J., Foucher, E., Robin, J.P., 2006. The English Channel stock of *Sepia officinalis*: modelling variability in abundance and impact of the fishery. Fish. Res. 78, 96–106.

Rumrill, S.S., 1990. Natural mortality of marine invertebrate larvae. Ophelia 32, 163–198.

Sauer, W.H.H., Smale, M.J., Lipinski, M.R., 1992. The location of spawning grounds, spawning and schooling behaviour of the squid *Loligo vulgaris reynaudii* (Cephalopoda: Myopsida) off the Eastern Cape Coast, South Africa. Mar. Biol. 114, 97–107.

Sauer, W.H.H., Lipinski, M.R., Augustyn, C.J., 2000. Tag recapture studies of the chokka squid *Loligo vulgaris reynaudii* d'Orbigny, 1845 on inshore spawning grounds on the south-east coast of South Africa. Fish. Res. 45, 283–289.

Segawa, S., 1990. Food consumption, food conversion and growth rates of the oval squid *Sepioteuthis lessoniana* by laboratory experiments. Nippon Suisan Gakk. 56, 217–222.

Shashar, N., Rutledge, P.S., Cronin, T.W., 1996. Polarization vision in cuttlefish—a concealed communication channel? J. Exp. Biol. 199, 2077–2084.

Shohet, A.J., Baddeley, R.J., Anderson, J.C., Kelman, E.J., Osorio, D., 2006. Cuttlefish responses to visual orientation of substrates, water flow and a model of motion camouflage. J. Exp. Biol. 209, 4717–4723.

Sobrino, I., Silva, L., Bellido, J.M., Ramos, F., 2002. Rainfall, river discharges and sea temperature as factors affecting abundance of two coastal benthic cephalopod species in the Gulf of Cadiz (SW Spain). Bull. Mar. Sci. 71, 851–865.

Sogard, S.M., 1997. Size-selective mortality in the juvenile stage of teleost fishes: a review. Bull. Mar. Sci. 60, 1129–1157.

Sollberger, A., 1965. Biological Rhythm Research. Elsevier Science Ltd, Amsterdam.

Staver, J.M., Strathmann, R.R., 2002. Evolution of fast development of planktonic embryos to early swimming. Biol. Bull. 203, 58–69.

Steer, M.A., Pecl, G.T., Moltschaniwskyj, N.A., 2003. Are bigger calamary *Sepioteuthis australis* hatchlings more likely to survive? A study based on statolith dimensions. Mar. Ecol. Prog. Ser. 261, 175–182.

Steer, M.A., Moltschaniwskyj, N.A., Nichols, D.S., Miller, M., 2004. The role of temperature and maternal ration in embryo survival: using the dumpling squid *Euprymna tasmanica* as a model. J. Exp. Mar. Biol. Ecol. 307, 73–89.

Strathmann, R.R., 1985. Feeding and nonfeeding larval development and life-history evolution in marine invertebrates. Annu. Rev. Ecol. Syst. 16, 339–361.

Strathmann, R.R., 2007. Three functionally distinct kinds of pelagic development. Bull. Mar. Sci. 81, 167–179.

Strathmann, R.R., Chaffee, C., 1984. Constraints on egg masses. II. Effect of spacing, size, and number of eggs on ventilation of masses of embryos in jelly, adherent groups, or thin-walled capsules. J. Exp. Mar. Biol. Ecol. 84, 85–93.

Strathmann, R.R., Strathmann, M.F., 1995. Oxygen supply and limits on aggregation of embryos. J. Mar. Biol. Assoc. U.K. 75, 413–428.

Sykes, A.V., Almansa, E., Lorenzo, A., Andrade, J.P., 2009. Lipid characterization of both wild and cultured eggs of cuttlefish (*Sepia officinalis* L.) throughout the embryonic development. Aquacult. Nutr. 15, 38–53.

Tsui, M.T.K., Wang, W.X., 2007. Biokinetics and tolerance development of toxic metals in *Daphnia magna*. Environ. Toxicol. Chem. 26, 1023–1032.

Unrine, J.M., Jackson, B.P., Hopkins, W.A., Romanek, C., 2006. Isolation and partial characterization of proteins involved in maternal transfer of selenium in the western fence lizard (*Sceloporus occidentalis*). Environ. Toxicol. Chem. 25, 1864–1867.

Vézina, A.F., Hoegh-Guldberg, O., 2008. Effects of ocean acidification on marine ecosystems. Mar. Ecol. Prog. Ser. 373, 191–201.

Vidal, E.A.G., Dimarco, F.P., Wormuth, J.H., Lee, P.G., 2002. Influence of temperature and food availability on survival, growth and yolk utilization in hatchling squid. Bull. Mar. Sci. 71, 915–931.

Villanueva, R., Moltschaniwskyj, N.A., Bozzano, A., 2007. Abiotic influences on embryo growth: statoliths as experimental tools in the squid early life history. Rev. Fish Biol. Fish. 17, 101–110.

Voss, G.L., 1983. A review of cephalopod fisheries biology. Mem. Natl. Mus. Vict. 44, 229–241.

Walsh, L.S., Turk, P.E., Forsythe, J.W., Lee, P.G., 2002. Mariculture of the loliginid squid *Sepioteuthis lessoniana* through seven successive generations. Aquaculture 212, 245–262.

Waluda, C.M., Trathan, P.N., Rodhouse, P.G., 1999. Influence of oceanographic variability on recruitment in the *Illex argentinus* (Cephalopoda: Ommastrephidae) fishery in the South Atlantic. Mar. Ecol. Prog. Ser. 183, 159–167.

Wang, J., Pierce, G.J., Boyle, P.R., Denis, V., Robin, J.P., Bellido, J.M., 2003. Spatial and temporal patterns of cuttlefish (*Sepia officinalis*) abundance and environmental influences-a case study using trawl fishery data in French Atlantic coastal, English Channel, and adjacent waters. ICES J. Mar. Sci. 60, 1149.

Ward, P.D., Boletzky, S., 1984. Shell implosion depth and implosion morphologies in three species of Sepia (Cephalopoda) from the Mediterranean Sea. J. Mar. Biol. Assoc. U.K. 64, 955–966.

Warnau, M., Temara, A., Jangoux, M., Dubois, P., Iaccarino, M., de Biase, A., Pagano, G., 1996. Spermiotoxicity and embryotoxicity of heavy metals in the echinoid *Paracentrotus lividus*. Environ. Toxicol. Chem. 15, 1931–1936.

Wells, M.J., 1958. Factors affecting reactions to Mysis by newly hatched Sepia. Behaviour 13, 96–111.

Wells, M.J., Wells, J., 1970. Observations on the feeding, growth rate and habits of newly settled *Octopus cyanea*. J. Zool. 161, 65–74.

Williamson, R., 1995. A Sensory Basis for Orientation in Cephalopods. J. Mar. Biol. Assoc. U.K. 75, 83–92.

Wolfram, K., Mark, F.C., John, U., Lucassen, M., Portner, H.O., 2006. Microsatellite DNA variation indicates low levels of genetic differentiation among cuttlefish (*Sepia officinalis* L.) populations in the English Channel and the Bay of Biscay. Comp. Biochem. Physiol. D Genom. Proteom. 1, 375–383.

Zeebe, R.E., Wolf-Gladrow, D., 2001. CO_2 in Seawater: Equilibrium, Kinetics, Isotopes. Elsevier Science BV, Amsterdam.

Zouhiri, S., Vallet, C., Mouny, P., Dauvin, J.C., 1998. Spatial distribution and biological rhythms of suprabenthic mysids from the English Channel. J. Mar. Biol. Assoc. U.K. 78, 1181–1202.

CHAPTER TWO

A Systematic Review of Phenotypic Plasticity in Marine Invertebrate and Plant Systems

Dianna K. Padilla[1], Monique M. Savedo

Department of Ecology and Evolution, Stony Brook University, Stony Brook, New York, USA
[1]Corresponding author: e-mail address: dianna.padilla@stonybrook.edu

Contents

Abstract

Marine organisms provide some of the most important examples of phenotypic plasticity to date. We conducted a systematic review to cast a wide net through the literature to examine general patterns among marine taxa and to identify gaps in our knowledge. Unlike terrestrial systems, most studies of plasticity are on animals and fewer on plants and algae. For invertebrates, twice as many studies are on mobile than sessile species and for both animals and plants most species are benthic intertidal zone taxa. For invertebrates, morphological plasticity is most common, while chemical plasticity is most common among algae. For algae, as expected, predators (inducible defences) are the primary cue for triggering plasticity. Surprisingly, for invertebrates the abiotic environment is the most common trigger for plasticity. Inducible defences in invertebrates have received great attention and predominate for a few well-studied species, which can bias perceptions; but, their predominance overall is not supported by the full data set. We also identified important research needs, including the need for data on non-temperate zone taxa, planned experiments to directly test the role of habitat variability

Advances in Marine Biology, Volume 65
ISSN 0065-2881
http://dx.doi.org/10.1016/B978-0-12-410498-3.00002-1

and the prevalence of plasticity. We also need information on the lag time for induction of plastic traits, which is critical for determining the adaptive value of phenotypic plasticity. Studies of early life stages and studies that link plasticity to mechanisms that produce phenotypes are critically needed, as are phylogenetic comparative studies that can be used to examine responses of organisms to both short- and long-term change.

Keywords: Phenotypic plasticity, Inducible defences, Inducible offences, Marine invertebrates, Marine algae

1. INTRODUCTION

Phenotypic plasticity, or the ability of a single genotype to produce different phenotypes, has captivated the interest of organismal biologists for decades, since early works by Levins (1963, 1968) and Bradshaw (1965). Much of the early literature on phenotypic plasticity focused on terrestrial systems, especially plants and insects (but see Woltereck, 1909, cited in Gilbert, 2012). Aquatic systems, including marine organisms, have since played an important role in the study of phenotypic plasticity, especially plasticity that is believed to be adaptive (Woltereck, 1909), including inducible defences (e.g. reviewed in Harvell, 1990a,b; Tollrian and Harvell, 1999). Phenotypic plasticity has been considered especially important ecologically when environmental change happens over relatively short periods of time or is spatially patchy, and different phenotypes have markedly different performance under different conditions (Miner et al., 2005). The adaptive value of plasticity can be limited if the time course of the response of the organism is longer than the frequency of environmental change (Padilla and Adolph, 1996), and thus may be particularly important for sessile or slow moving species.

Understanding the mechanisms that control and lead to—or allow—phenotypic plasticity has been tied to developmental biology and the evolution of phenotypes. Phenotypic plasticity is central to several Grand Challenge questions in organismal biology, including understanding the links between genotypes and phenotypes, and how organisms maintain their function and performance, as well as adaptability in the face of environmental change (Mykles et al., 2010; Schwenk et al., 2009). Phenotypic plasticity has played a central role in studies of evolutionary responses to changing environments, including the evolution of diversity (West-Eberhard, 1989, 2003). There is also great interest in the ecological implications of plasticity, both for individual species (Miner et al., 2005) and for potential cascading effects of organismal plasticity and phenotypically plastic responses in natural communities (Agrawal, 2001; Miner et al., 2012; Wade, 2007). In marine systems, there is increasing interest in the role of phenotypic plasticity in species interactions, including predator–prey interactions (sometimes referred to as trait-mediated-interactions, or TMI, or non-consumptive effects; e.g. Gimenez, 2004). Phenotypic plasticity is also increasingly of interest in pressing environmental issues such as invasion by non-native species, both for the identification of traits that make species good invaders and for the impacts that introduced species have on communities they invade (Sih et al., 2010).

The adaptive value and evolution of phenotypic plasticity has generated decades of research on a variety of taxa and has led to interest in whether phenotypic plasticity *per se* can be a mechanism for evolution and lead to taxonomic diversification (e.g. West-Eberhard, 1989, 2003). In most cases, the likelihood that phenotypic plasticity will be adaptive depends on many factors. These include costs and limits of constitutive versus plastic phenotypes (Auld et al., 2010; DeWitt et al., 1998; Dodson, 1984; Tollrian, 1995), including opportunity costs, such as performance differences when phenotypes are matched versus mismatched to their environments, or when there is a lag time between environmental change (or detection of environmental change) and the production of a new phenotype (Padilla and Adolph, 1996). The reliability of cues that induce new phenotypes can also be an important limit to the adaptive value of a phenotypic plasticity (DeWitt et al., 1998). A critical component of any adaptive plasticity is the ability of organisms to correctly assess environmental conditions and recognize cues that correspond to changes in both the abiotic and biotic environment (Bourdeau, 2010; Piersma and Drent, 2003; Reed et al., 2010).

Marine organisms have provided many important examples of phenotypic plasticity and have been the focus of experimental studies of plasticity for decades. An incredibly wide range of traits of marine organisms have been shown to be phenotypically plastic, including those that are behavioural, morphological, physiological, biochemical, chemical, or related to life history. Plasticities can result from responses to the environment during the lifetime of an individual (reviewed in Adams et al., 2011; Miner et al., 2005) or across generations. Clonal animals that show such cross-generational changes include model systems for the study of plasticity like Cladocera (Dodson, 1981; Tollrain and Dodson, 1999; Tollrian, 1993, 1995) and aphids (Moran, 1992), but are more rare in marine systems. An exception is seen in bryozoans, such as *Membranipora membranacea*, where individual zooids cannot change morphology, but when a colony is exposed to cues from predators, new zooids are produced with defensive spines (Harvell, 1984, 1986). In some cases, changes in morphological traits can be produced by direct feedback between a behaviour or environmental condition and the modulation of an existing morphology, as is seen in use-induced traits, for example, crabs produce stouter claws when feeding on harder prey (Smith and Palmer, 1994) or fish develop wider, more robust jaws when feeding on hard shelled prey (e.g. Frederich et al., 2012). Recently, there has been increased interest in the role of behavioural plasticity (Luttbeg and Sih, 2010) and how it may be linked to morphological plasticity, especially inducible morphological defences (Bourdeau and Johansson, 2012).

Marine systems have provided some of best cases of experimental demonstrations of adaptive inducible defences and life history plasticity. However, gaps in our understanding of the generality of types of plasticity, their prevalence among different taxa or different habitats, and the evolutionary consequences of such plasticities remain. Other reviews have delved into the potential ecological and evolutionary significance of phenotypic plasticities in marine systems (e.g. Dudgeon and Kübler, 2011; Harvell, 1990a,b). Our goal was to take a different approach, and cast a wide net through the published literature by using a systematic review and determine if general patterns among marine taxa exist regarding plasticity, as well as identify holes in our knowledge.

We had several questions we wanted to address. First, we wanted to determine if phenotypically plastic traits are common in general, and if they are equally common among different taxa, among taxa in different geographic regions, and among taxa in different habitat types. Second, we wanted to know whether plasticity was more common for different types of traits (e.g. morphological, behavioural, physiological or chemical traits) and, third, whether traits important for organismal defences from consumers, offences (traits affecting competitive ability or the ability to consume prey) or responses to the abiotic environment were more common. We also wanted to know if patterns of phenotypic plasticity in marine plants (including algae) were similar to those of marine invertebrates.

1.1. Why a systematic review

A systematic review is intended to be an exhaustive review of the published literature focused on a research question. The goal is generally to identify, through repeatable search methods, and synthesize all high quality research evidence relevant to that question (Pullin and Stewart, 2006). This methodology uses a transparent, repeatable approach to review studies and minimizes bias. Thus, a systematic review allows us to survey a very broad range of the literature, and not limit our findings to literature already most familiar to us. This type of survey approach will allow us to detect patterns that might not otherwise be seen when focusing only on particular studies or systems that are most familiar. Such biases can obscure larger patters that are in the literature, thus this approach is more amenable to determining if generalizations are system or taxon specific.

At present, this synthesis tool is used primarily in applied fields, such as medicine, conservation and management (Pullin and Stewart, 2006). But, it is becoming more common in a range of scientific fields, and especially in conjunction with other types of synthetic research tools, such as meta-analysis.

The first step of any systematic review is a thorough, repeatable search of the literature for relevant papers. With the advent of electronically available literature databases, such as the Web of Science, such systematic reviews are more feasible. Once studies have been collected, they are then screened to make sure they fit the question of interest, and often the quality of the data included are examined or ranked. Data are then extracted from selected studies and synthesized in some fashion.

2. METHODS

We conducted a systematic review of the published literature searchable in the Web of Science from 1900 to January 2012 to find publications on phenotypic plasticity in marine invertebrate animals and marine algae and plants. This database does not include all scientific journals, and does not search books or many monographs, but facilitates searching a broad range of literature in a relatively unbiased fashion. To identify as much of the relevant literature available as possible, we used the following search terms:

marine AND phenotypic* plastic*, marine AND inducible,

inducible defen* AND marine, inducible offen* AND marine

We then read the titles and abstracts of collected literature and removed studies clearly not relevant (e.g. biomedical studies, studies of plastics and plastic pollution in marine systems), narrowing our findings to a total of 879 studies. Because of the large amount of literature, we also removed studies of fish and other vertebrates, reducing the list to 536 studies of invertebrates, algae and plants to consider further. To further narrow the literature

vast majority of studies that looked for plasticity in a trait found it. This result could mean that plasticity is very widespread and is present in the vast majority of species. If this is the case, more focus is needed on organisms and traits where plasticity is not found, as those cases may be the most informative regarding the limits or constraints on phenotypic plasticity. Alternatively, these results may be due to publication or study bias. Researchers may preferentially study species or traits of species where they have some evidence that plasticity is likely to be found and do not pursue those species or traits where plasticity is not likely. Or, researchers may decide not to publish results where phenotypic plasticity of traits was not found.

The first study explicitly addressing phenotypic plasticity in a marine organism found in our systematic review was published in 1984 (Harvell, 1984). There were relatively few publications over the next decade (under two publications per year). Then, publication rates in this area increased starting in 1996, averaging almost nine per year from 1996 to 2005. There was a continued increase in publication rate from 2006 to 2011 with just under 22 publications per year (Figure 2.1). Through time, there has been a slight increase in the number of studies that have tested for but not found plasticity (Figure 2.2). To date, the number of such studies still remains low, around three per year in recent years. Surprisingly, through time, there has been a relatively constant rate of the publication of reviews (Figure 2.3).

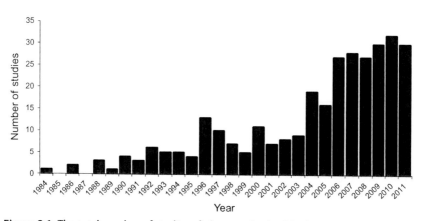

Figure 2.1 The total number of studies of phenotypic plasticity in marine invertebrates and plants found in our systematic review by year. We searched the literature from 1900 to January 2012. Data are presented through December 2011. The first paper that fit our search criteria was published in 1984.

Studies where plasticity was not found

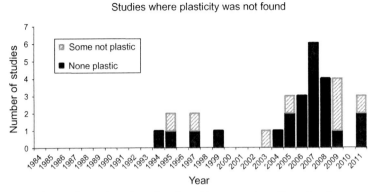

Figure 2.2 The total number of studies found in our systematic review by year where phenotypic plasticity was tested for but not found (solid bars). The stacked hatched bars are additional studies where some of the species tested displayed phenotypic plasticity, while for other species tested no phenotypic plasticity was found. We searched the literature from 1900 to January 2012. Data are presented through December 2011.

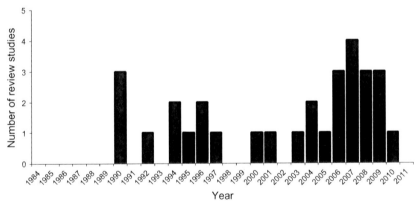

Figure 2.3 The total number of review papers on phenotypic plasticity in marine invertebrates and plants found in our systematic review by year. We searched the literature from 1900 to January 2012. Data are presented through December 2011. The first review papers found were published in 1990.

3.1. Diversity of studies and taxa

The vast majority of studies (94.6%) were based on experimental data, either laboratory experiments (79.8%, 192 studies with animals, 36 studies with algae) or field experiments (14.8%, 35 with animals, 5 with algae; Table 2.1). 21.5% of studies included field observational data, either alone

Table 2.1 The number of studies of phenotypic plasticity of marine invertebrate animals and plants (vascular plants, algae and fungi) that were experimental or observational and the life stage studied

Phylum	Studies	Data source				Life stage studied		
		Lab experiment	Field experiment	Lab observation	Field observation	Embryo/ larval	Juvenile	Adult
Animal								
Annelida	6	4	0	1	2	2	0	3
Arthropoda	40	23	7	1	15	17	14	26
Bryozoa	21	20	5	0	3	6	6	20
Chordata	1	1	0	0	0	0	0	1
Cnidaria	16	14	3	1	5	2	4	14
Enchinodermata	37	35	1	0	1	33	6	11
Mollusca	118	89	18	2	26	19	20	94
Porifera	6	2	0	0	0	0	0	2
Platyhelminthes	2	4	1	1	3	0	0	6
Total		192	35	6	55	79	50	177

Plant

Plant						
Chlorophyta	4	1	4	5	0	3
Fungi	2	0	0	0	0	2
Haptophyta	1	0	1	0	0	1
Heterokontophyta	18	4	0	1	0	27
Myzozoa	3	0	0	0	0	3
Planta	0	0	0	1	0	1
Rhodophyta	8	0	1	0	1	8
Total	36	5	6	7	1	45

Some studies included more than one type of data (e.g. both experimental and observational or both laboratory and field experimental) and multiple life stages.

or in combination with experimental data, and lab observational data were recorded for 3.7% of studies. Thus, the majority of the studies were based on relatively higher quality data (experimental results or experimental and observational data combined) rather than just observational data. The patterns of findings from observational studies were usually the same as for the experimental studies, so they do not appear to have presented a bias in our findings. Most studies were also on adults, rather than embryos or larvae or juveniles. This was most extreme for algae, where there was only a single study that focused on juveniles (Santelices and Varela, 1993), and the rest were on adult life stages. For animals, 25.8% of studies focused on embryos or larval stages, 16.3% on juveniles and 57.8% of studies on adult stages (Table 2.1).

Unlike terrestrial systems, most studies of plasticity were on animals and only 17.6% were studies of vascular plants (1 species, 1 study), fungi (2 species, 1 study) or algae (58 total) (hereafter all referred to as plants for simplicity; Table 2.2). Surprisingly, for both animals and plants, 90% of studies found some sort of plasticity. The diversity of taxa studied was not huge. For animals, only 9 (of over 40) phyla, 20 total classes and 138 species total were studied. Studies of vascular plants (1 species), fungi (2 species) and algae (25 species), included 7 phyla, 9 classes and 38 total species (Table 2.2).

For animals, twice as many studies were of mobile species (67%), than sessile or sedentary species (33%). This result is not due to the inclusion of behavioural plasticity in our study. For the majority of species where behavioural plasticity was found, plasticity in other traits (morphological, physiological, chemical or life history) was also found. And, in some cases, behavioural plasticity was found in sessile animals (e.g. feeding behaviour). For plants, 86.3% were of sessile species and only 13.7% were of mobile species (phytoplankton). Thus, the generalization that plasticity is more common among sessile than mobile organisms is not supported for marine invertebrates, a conclusion that would be difficult to make without testing a wide range of taxa.

3.2. Habitat and geography

Not surprisingly, the vast majority of studies for both animals and algae were on benthic organisms (83%), and of those, a very large percentage of organisms are found in the intertidal zone (Table 2.3). For animals, the majority of pelagic organisms studied (>80%) were the larval stages of benthic

Table 2.2 The diversity of taxa of marine invertebrate animals and plants (vascular plants, algae and fungi) that were the focus of studies of phenotypic plasticity

Phylum	Class	# Species	# Genera	Total studies
Animals				
Annelida	Polychaeta	4	4	6
Arthropoda	Total	31	27	40
	Crustacea Malacostraca Amphipoda	2	1	2
	Crustacea Malacostraca Decapoda	13	11	15
	Crustacea Malacostraca Stomatopoda	1	1	1
	Crustacea Maxillopoda Cirripedia	6	4	12
	Crustacea Maxillopoda Copepoda	7	7	8
	Crustacea Ostracoda	2	2	2
Bryozoa	Gymnolaemata	4	4	22
Chordata	Urochordata	1	1	1
Cnidaria	Total	15	14	16
	Anthozoa	12	11	13
	Hydrozoa	2	2	2
	Scyphozoa	1	1	1
Echinodermata	Total	18	14	37
	Asteriodea	2	2	2
	Echinoidea	12	9	31
	Holothuroidea	1	1	1
	Ophiuroidea	3	2	3
Mollusca	Total	58	37	117
	Bivalvia	13	11	25
	Gastropoda	45	26	92

Continued

Table 2.2 The diversity of taxa of marine invertebrate animals and plants (vascular plants, algae and fungi) that were the focus of studies of phenotypic plasticity—cont'd

Phylum	Class	# Species	# Genera	Total studies
Platyhelminthes	Turbellaria	1	1	2
Porifera	Calcarea	6	6	6
Total: 9	20	138	94	231
Plants				
Chlorophyta	Total	8	4	8
	Bryopsidophyceae	7	3	7
	Ulvophyceae	1	1	1
Fungi	Mitosporic Fungi	2	1	1
Haptophyta	Prymnesiophyceae	1	1	1
Heterokontophyta	Total	17	12	28
	Bacillariophyceae	4	4	4
	Phaeophyceae	13	8	24
Myzozoa	Dinophyceae	3	2	3
Planta	Angiosperm	1	1	1
Rhodophyta	Florideophyceae	6	6	8
Total: 7	9	38	27	50

For most phyla, the number of studies per class is also provided. If studies included members of more than one class, the phylum totals are provided for the number of species and genera studied, and the total number of studies found in our systematic review. For the Arthropoda, all studies were on taxa in the subphylum Crustacea. Class is provided, and for some groups, order is also provided.

organisms. For both animals and plants, $\sim$57% of benthic studies focused on intertidal zone organisms. While we may predict that the environmental variability seen by intertidal organisms may make plasticity more important, until we have more studies of subtidal species, and have enough studies where plasticity is tested for but not found, it is difficult to know whether these findings are due to real patterns or low sampling of subtidal or pelagic species (Table 2.3). The vast majority of studies (77%) have been conducted on temperate taxa. Of the remainder, 8% have been on tropical or subtropical taxa and only 1% on Arctic or Antarctic taxa. Only one study involved a deep sea species (Jollivet et al., 1995).

Table 2.3 The number of studies of phenotypic plasticity in organisms from different geographic regions and habitat types

Phylum	Studies	Region				Habitat		
		Tropical	Subtropical	Temperate	Arctic/ Antarctic	Pelagic	Benthic all	Benthic intertidal
Animals								
Annelida	6	0	0	5	0	2	6	2
Arthropoda	40	7	4	29	0	12	28	17
Bryozoa	21	0	0	22	0	1	20	0
Chordata	1	0	0	1	0	0	1	0
Cnidaria	16	9	1	6	0	1	14	2
Enchinodermata	37	2	7	28	0	21	20	4
Mollusca	118	4	9	102	3	5	97	85
Porifera	6	6	0	0	0	0	2	0
Platyhelminthes	2	0	0	2	0	1	6	0
Total		28	21	195	3	43	194	110

Continued

Table 2.3 The number of studies of phenotypic plasticity in organisms from different geographic regions and habitat types—cont'd

Phylum	Studies	Region				Habitat		
		Tropical	Subtropical	Temperate	Arctic/ Antarctic	Pelagic	Benthic all	Benthic intertidal
Plants								
Chlorophyta	8	5	0	3	0	0	4	3
Fungi	2	0	2	2	0	0	2	0
Haptophyta	1	0	0	1	0	1	0	0
Heterokontophyta	28	3	2	23	0	3	25	16
Myzozoa	3	0	0	3	0	3	0	0
Planta	1	1	0	0	0	0	1	0
Rhodophyta	8	1	0	7	0	0	8	4
Total		10	4	39	0	7	40	23

For animals, there was also one species that was studied in the deep sea. For benthic organisms, we also determined the number of studies of species that are found primarily in the intertidal zone.

3.3. Types of responses

For invertebrate marine animals, morphological plasticity was the most common form of plasticity, found in 40% of studies, followed by behavioural plasticity (23.5%), life history plasticity (20%) and physiological plasticity (14%; Table 2.4). Few animals were found to have plasticity in chemical phenotypes (2.5%). The one vascular plant that was studied, the seagrass *Thalassia testudinum*, showed morphological plasticity (Bricker et al., 2011), and the two species of fungi studied (*Dendryphiella arenaria* and *Dendryphiella salina*; de la Cruz et al., 2006) were found to have physiological plasticity. Among the algae, 46.8% showed morphological plasticity and 44.7% showed chemical plasticity, much more than that seen in animals. 6.4% of algae were found to have some form of physiological plasticity, and one species (2.1%) displayed life history plasticity.

Most attention in the literature, especially in reviews (e.g. Adler and Harvell, 1990; Harvell, 1990a,b; Pohnert et al., 2007; Toth and Pavia, 2007), has been paid to the role of inducible defences. Surprisingly, however, they did not dominate our findings for marine animals. Our impression of their dominance may be influenced by the large number of studies that have been published on certain taxa that display inducible defences (see below). We found that for marine animals the most common documented cases of plasticity were in response to the abiotic environment (42.4%; Table 2.4). This was followed by inducible defences (responses to consumers, 37.4%) and inducible offences (responses to competitors 6.7% or prey 13.4%, 20.1% total; Table 2.4). This last finding was also not expected as recent studies have concluded that inducible offences are much more rare in nature than inducible defences (Mougi et al., 2011).

For algae, as expected, inducible defences were the most common form of plasticity (58%), and the rest of the observed plasticity was in response to the environment (42%). For the one plant and two species of fungi studied, the observed plasticity was in response to the physical environment (Table 2.4).

3.4. Influence of most-studied systems

Of the total 176 species found through our review, the majority, 133 were the subject of only a single study. Thus, we asked the question of whether our impressions of the prevalence of certain types of plasticity may be driven by a few well-studied taxa. 37% of all studies were on molluscs, primarily gastropods (92 total studies; Table 2.2). Eleven species of animals and two

Table 2.4 The types of cues that triggered phenotypically plastic responses and the types of traits that were phenotypically plastic

Phylum	Studies	Cue				Type of response				
		Consum/ Enemy	Comp	Prey	Environ	Morpho	Chem	Physio	Life history	Behav
Animals										
Annelida	6	1	0	0	3	0	1	2	3	3
Arthropoda	40	10	3	9	21	19	0	7	12	17
Bryozoa	21	9	6	0	8	16	1	5	9	6
Chordata	1	0	1	0	0	1	0	0	1	1
Cnidaria	16	6	4	0	9	10	3	4	2	5
Enchinodermata	37	6	0	14	10	27	0	12	14	10
Mollusca	118	56	0	8	47	70	1	19	29	43
Porifera	6	0	2	0	0	3	3	0	0	0
Platyhelminthes	2	1	0	0	4	0	0	1	2	0
Total		89	16	31	102	146	9	50	72	85

Plants

	Total	consum/enemy	comp	environ	morph	chem	physio	life history	behav
Chlorophyta	8	3	0	0	0	6	3	1	0
Fungi	2	0	0	0	2	0	0	2	0
Haptophyta	1	1	0	0	0	1	0	0	0
Heterokontophyta	28	14	0	0	8	7	13	2	1
Myzozoa	3	3	0	0	1	2	1	0	0
Planta	1	0	0	0	1	1	0	0	0
Rhodophyta	8	4	0	0	3	6	4	0	0
Total	51	25	0	0	15	23	21	5	1

Cues included consumers or enemies (consum/enemy), competitors (comp), prey or abiotic environmental factors (environ). Traits that were plastic including morphological (morph), chemical or biochemical (chem), physiological (physio), life history, or behavioural (behav) traits.

species of algae were the subject of five or more studies. The most-studied species were animals: gastropods *Littorina obtusata* (13 studies), *Nucella lapillus* (9), *Nucella lamellosa* (9), the bryozoan *Membranipora membranacea* (13) and the bivalve *Mytilus edulis* (8). Other important species were the snail *Littorina saxatilis*, the bryozoan *Bugula neritina*, the urchin *Strongylocentrotus droebachiensis* and the brown alga *Ascophyllum nodosum*, each with six studies, and the gastropod *Littorina littorea*, sand dollar *Dentraster excentricus*, barnacle *Balanus glandula* and brown rockweed *Fucus vesiculosus*, each with five studies.

All of the five most-studied animals were temperate benthic species, and four of the five live in the intertidal zone; only *Membranipora membranacea* is primarily subtidal. Morphological plasticity was the most common form of plasticity seen among these taxa (average of 77% of studies for each species). An average of 35% of studies were of behavioural plasticity, and 14% physiological and 11% on life history traits. An average of 72% of studies identified responses as inducible defences, and only 6% were inducible offences. An average of 27% found plastic responses to abiotic environmental conditions. Thus, focusing on the most-studied systems, it would appear that inducible defences are the most important form of phenotypic plasticity in marine animals, a conclusion not supported by the full data set for all species studied (Table 2.4). For molluscs as a whole, 51% of studied plastic responses were inducible defences, 8% were inducible offences and 41% were in response to the abiotic environment. Therefore, even for molluscs, the most-studied species paint a different picture of the relative importance of different types of phenotypically plastic responses.

Two brown algae, *A. nodosum* and *F. vesiculosus*, were the most frequent subjects of studies of plasticity among plants. For *A. nodosum*, there were six studies that found plasticity in inducible chemistry in response to consumers or to UV irradiation (Pavia and Brock, 2000; Pavia and Toth, 2000; Pavia et al., 1997; Svensson et al., 2007), but one that found no phenotypically plastic response to herbivory (Long and Trussell, 2007). For *F. vesiculosus*, all five studies found induced chemical defences in response to consumers (Haavisto et al., 2010; Peckol et al., 1996; Rohde and Wahl, 2008; Rohde et al., 2004), and one study also found responses to the physical environment (Peckol et al., 1996).

3.5. Research needs

The value of a systematic review is not only that it allows for the detection of patterns that might not otherwise be seen by focusing only on particular studies or systems that are most familiar, but also because it can be used to find areas where more research is needed to address questions of interest. From our review, it is clear that although there has been extensive research on some taxa, the diversity of taxa for which we do not have information is great, limiting our ability to draw robust conclusions. Some research needs are especially critical to answer pressing questions regarding the ecology and evolution of phenotypic plasticity.

3.5.1 Is plasticity extremely common or do we lack sufficient data on taxa that do not show plasticity?

Much of the theoretical research on phenotypic plasticity has focused on conditions that would favour plasticity, and limits and costs of phenotypic plasticity that would affect plasticity as an adaptive strategy (Auld et al., 2010; DeWitt and Scheiner, 2004; DeWitt et al., 1998; Padilla and Adolph, 1996; Relyea, 2002; Scheiner, 1993). More empirical studies designed to test predictions of theoretical models would help answer this question, especially focused research on the traits and taxa that are not predicted to be plastic. Evolutionary patterns within clades of organisms are best detected when there is a well-resolved phylogeny that can be used in conjunction with information about a large number of taxa and statistical phylogenetic comparative methods can be used in conjunction with well-designed experiments (Weber and Agrawal, 2012).

The overwhelming conclusion of our review is that phenotypic plasticity is very common. However, we need more studies of species where plasticity is not suspected and more reports of findings when plasticity is tested for but not found, to be sure that our conclusions are not based on publication bias. Certainly, there are ecological conditions when fixed phenotypes would be advantageous or selected for (e.g. constitutive defences when consumers are reliably continuously abundant). Focused work on clades with well-resolved phylogenies and with many species that are amenable to direct experimentation, but predicted to be more or less phenotypically plastic, would be ideal. Such a work would also be important for answering the larger question of whether phenotypic plasticity has affected evolutionary diversification, either enhancing or retarding the rate of speciation within clades (West-Eberhard, 2003).

3.5.2 What are the time lags between detection and phenotypic response, and do they limit the adaptive value of plasticity?

The lag time between when an organism experiences a new environment or receives cues that trigger a phenotypic switch and the realization of the new phenotype has been identified as an important factor that can limit the adaptive value of plastic phenotypes (Padilla and Adolph, 1996). This lag time is especially important if it is longer than the rate at which the organisms experiences environmental change, or there is a very large fitness cost for phenotypes mismatched to their environment.

Surprisingly, only about 10% of studies in our systematic review reported the lag time for the induction of a new phenotype. These lag times ranged from less than 1 h (e.g. Paul and Van Alstyne, 1992) to 1–2 days (e.g. Harvell and Padilla, 1990), to 3–4 weeks (e.g. morphological trait, Hadfield and Strathmann, 1996; Padilla, 2001), to more than 6 months (e.g. Neo and Todd, 2011). For the few reports of lag times that we found, there was no tendency for behavioural traits to have shorter lag times than morphological or chemical traits. Chemical traits could respond in less than 1 h (Paul and Van Alstyne, 1992), and some morphological traits were also rapidly deployed in 1–2 days (Harvell and Padilla, 1990), while the lag time for some behavioural traits was reported to be 5 days or more (Calado et al., 2010). Without more information on lag times, we will not be able to determine if certain traits have longer lag times than others, or the real-world adaptive value of phenotypically plastic traits that we observe.

3.5.3 Is plasticity more common in variable environments?

Marine systems in temperate zones are thought to be more variable in terms of physical environmental factors (e.g. temperature, waves, storms) leading to greater diversity in the tropics, where more intense biological interactions such as predation influence species (Vermeij, 1978, 1987). Most studies were conducted on temperate benthic species, primarily those found in the intertidal zone, making it difficult to determine if our findings are robust or the consequence of publication/study bias. Most scientific institutions and research labs are in temperate regions, and benthic intertidal zone species are very amenable to experimental studies. We need many more studies on organisms in both the tropics, and Arctic and Antarctic regions to determine if organisms in those regions are equally likely to display phenotypic plasticity. Within the temperate zone, more studies that explicitly address spatial and temporal variability in exposure to the cues that trigger a particular phenotypic plasticity would also be helpful for addressing this question. Vermeij (1987, 1994) has argued that for benthic marine species the risk of predation in the tropics is greater and has been so through evolutionary time, leading to escalations in anti-predator defences in shelled prey. Thus, constitutive defences may be more likely in tropical taxa than temperate ones, which may be more likely to display phenotypic plasticity.

3.5.4 Is plasticity more common in early life stages where developmental programs may be less fixed?

Most studies in our review were on adult organisms. For invertebrates, there were several studies on larval stages, but studies on juveniles were very rare (Table 2.1). For plants, this was more extreme, where there was only a single study on early life stages. Clearly, more data are needed on the plasticity of early life stages, and whether patterns of phenotypic plasticity seen in adults are also seen earlier in ontogeny.

Understanding the developmental mechanisms that underlie plasticity are critical for understanding the mapping between phenotype and genotype (e.g. Charrier et al., 2012), and the ability of organisms to accommodate both short- and long-term change, both important Grand Challenge questions for organismal biology (Schwenk et al., 2009). Thus, more study of early life stages, and especially studies that link plasticity to the mechanisms that control and produce phenotypes, will be critical. However, there is still a large disconnect between these two areas of science that needs to be bridged.

3.5.5 What are the long-term evolutionary consequences and patterns of phenotypic plasticity?

Ecological responses in the short term have been the focus of most studies of phenotypic plasticity, and detailed ecological consequences and patterns of such phenotypic flexibility are being studied, but primarily in a handful of systems. In order to understand the generality of such mechanisms, we need more studies across taxa, studies that are comparative across taxa and mechanisms, and responses of organisms to both short- and long-term change.

Both inducible offences and defences are important for ecological systems, affecting community structure and food web dynamics (e.g. Mougi and Kishida, 2009; Peacor et al., 2012). And, both can lead to coevolutionary feedbacks between consumers and prey (Kishida et al., 2011; Mougi, 2012; Mougi and Iwasa, 2011; Mougi and Kishida, 2009). Although it has been reported that inducible offences are more rare in nature than inducible defences (Mougi et al., 2011), we found that inducible offences are more common than expected for marine invertebrates.

For some taxa, direct, physical interactions are needed to trigger inducible defensive or offensive traits (e.g. Paul and Van Alstyne, 1992). For others, especially animal-induced defences, chemical signalling appears to be sufficient to induce phenotypically plastic defences (e.g. Bourdeau, 2009; Harvell, 1986) and may also trigger inducible offences (e.g. Padilla, 2001). The use of chemical signals to induce defences or offences can be used to detect consumers (or prey) before actual contact, allowing organisms to respond more quickly to changes in the environment (Barrett and Heil, 2012). By using chemical signals, prey could mount defences prior to attack by consumers and consumers could produce phenotypes most effective for particular prey. This could greatly reduce the lag time between detecting a predator or prey and producing a new phenotype, thereby reducing the opportunity costs and increasing the adaptive value of a phenotypic plasticity (Gabriel, 2005; Gabriel et al., 2005; Padilla and Adolph, 1996). Thus, the use of chemical cues to induce phenotypically plastic responses could reduce or prevent fitness losses in prey that result from predator damage and fitness losses in predators due to having less effective offensive phenotypes, and could thereby enhance predator–prey coevolution. More information about the use of chemical signals or other indirect means of assessing environmental change is needed to determine if plasticity has played an important role in the evolution of taxa. And, as mentioned above, well-resolved phylogenies and experimental studies of multiple closely related taxa are tools needed to address questions not only about evolutionary patterns of plasticity, but whether plasticity has played an important role affecting evolution, including the evolution of diversification.

4. CONCLUSIONS

The systematic review approach that we used proved to be a very useful tool to address questions about the prevalence of phenotypic plasticity (among taxa, geographic regions and habitat types), whether certain traits are most commonly phenotypically plastic, and the most common environmental triggers for the induction of phenotypic plasticity in marine invertebrates and plants. Some of our present assumptions about the types of plasticity that are most common in marine invertebrate and plant systems likely stem from having many studies on a few taxa. Different patterns emerge when all of the literature is considered together. More studies across taxa and systems, and studies that are designed to target critical questions about phenotypic plasticity are needed, especially studies that link ecological processes, evolutionary responses and links to developmental programs and flexibility that can control and determine phenotypic plasticity. We have made tremendous progress in investigating and documenting phenotypic plasticity in marine organisms over the past three decades. Now is the time to begin to focus our efforts on using this knowledge to address big, pressing questions in biology where phenotypic plasticity can or does play a central role.

REFERENCES

Adams, D.K., Sewell, M.A., Angerer, R.C., Angerer, L.M., 2011. Rapid adaptation to food availability by a dopamine-mediated morphogenetic response. Nat. Commun. 2, 592. http://dx.doi.org/10.1038/ncomms1603.

Adler, F.R., Harvell, C.D., 1990. Inducible defenses, phenotypic variability and biotic environments. Trends Ecol. Evol. 5, 407–410.

Agrawal, A.A., 2001. Ecology-phenotypic plasticity in the interactions and evolution of species. Science 294, 321–326.

Auld, J.R., Agrawal, A.A., Relyea, R.A., 2010. Re-evaluating the costs and limits of adaptive phenotypic plasticity. Proc. R. Soc. B 277, 503–511.

Barrett, L.G., Heil, M., 2012. Unifying concepts and mechanisms in the specificity of plant-enemy interactions. Trends Plant Sci. 17, 282–292.

Bourdeau, P.E., 2009. Prioritized phenotypic responses to combined predators in a marine snail. Ecology 90, 1659–1669.

Bourdeau, P.E., 2010. An inducible morphological defence is a passive by-product of behaviour in a marine snail. Proc. R. Soc. B 277, 455–462.

Bourdeau, P.E., Johansson, F., 2012. Predator-induced morphological defences as by-products of prey behaviour: a review and prospectus. Oikos 121, 1175–1190.

Bradshaw, A.D., 1965. Evolutionary significance of phenotypic plasticity in plants. Adv. Genet. 13, 115–155.

Bricker, E., Waycott, M., Calladine, A., Zieman, J.C., 2011. High connectivity across environmental gradients and implications for phenotypic plasticity in a marine plant. Mar. Ecol. Prog. Ser. 423, 57–67.

Calado, R., Pimentel, T., Pochelon, P., Olaguer-Feliu, A.O., Queiroga, H., 2010. Effect of food deprivation in late larval development and early benthic life of temperate marine coastal and estuarine caridean shrimp. J. Exp. Mar. Biol. Ecol. 384, 107–112.

Charrier, B., Le Bail, A., de Reviers, B., 2012. Plant Proteus: brown algal morphological plasticity and underlying developmental mechanisms. Trends Plant Sci. 17, 468–477.

de la Cruz, T.E., Wagner, S., Schulz, B., 2006. Physiological responses of marine *Dendryphiella* species from different geographical locations. Mycol. Progress 5, 108–119.

DeWitt, T.J., Scheiner, S.M., 2004. Phenotypic Plasticity: Functional and Conceptual Approaches. Oxford University Press, Oxford, UK, 272 pp.

DeWitt, T.J., Sih, A., Wilson, D.S., 1998. Costs and limits of phenotypic plasticity. Trends Ecol. Evol. 13, 77–81.

Dodson, S.I., 1981. Morphological variation of *Daphnia pulex* Leydig (Crustacea, Cladocera) and related species from North America. Hydrobiologia 83, 101–114.

Dodson, S.I., 1984. Predation of *Heterocope septentrionalis* on two species of *Daphnia*: morphological defenses and their cost. Ecology 65, 1249–1257.

Dudgeon, S.R., Kübler, J.E., 2011. Hydrozoans and the shape of things to come. Adv. Mar. Biol. 59, 107–144.

Frederich, B., Liu, S.Y.V., Dai, C.F., 2012. Morphological and genetic divergences in a coral reef damselfish, *Pomacentrus coelestis*. Evol. Biol. 39, 359–370.

Gabriel, W., 2005. How stress selects for reversible phenotypic plasticity. J. Evol. Biol. 18, 873–883.

Gabriel, W., Luttbeg, B., Sih, A., Tollrian, R., 2005. Environmental tolerance, heterogeneity, and the evolution of reversible plastic responses. Am. Nat. 166, 339–353.

Gilbert, S.F., 2012. Ecological developmental biology: environmental signals for normal animal development. Evol. Dev. 14, 20–28.

Gimenez, L., 2004. Marine community ecology: importance of trait-mediated effects propagating through complex life cycles. Mar. Ecol. Prog. Ser. 283, 303–310.

Haavisto, F., Valikangas, T., Jormalainen, V., 2010. Induced resistance in a brown alga: phlorotannins, genotypic variation and fitness costs for the crustacean herbivore. Oecologia 162, 685–695.

Hadfield, M.G., Strathmann, M.F., 1996. Variability, flexibility and plasticity in life histories of marine invertebrates. Oceanol. Acta 19, 323–334.

Harvell, C.D., 1984. Predator-induced defense in a marine bryozoan. Science 224, 1357–1359.

Harvell, C.D., 1986. The ecology and evolution of inducible defenses in a marine bryozoan—cues, costs, and consequences. Am. Nat. 128, 810–823.

Harvell, C.D., 1990a. The evolution of inducible defense. Parasitology 100, S53–S61.

Harvell, C.D., 1990b. The ecology and evolution of inducible defenses. Q. Rev. Biol. 65, 323–340.

Harvell, C.D., Padilla, D.K., 1990. Inducible morphology, heterochrony, and size hierarchies in a colonial invertebrate monoculture. Proc. Natl. Acad. Sci. U.S.A. 87, 508–512.

Jollivet, D., Desbruyeres, D., Ladrat, C., Laubier, L., 1995. Evidence for differences in the allozyme thermostability of deep-sea hydrothermal vent polychaetes (Alvinellidae)—a possible selection by habitat. Mar. Ecol. Prog. Ser. 123, 125–136.

Kishida, O., Trussell, G.C., Ohno, A., Kuwano, S., Ikawa, T., Nishimura, K., 2011. Predation risk suppresses the positive feedback between size structure and cannibalism. J. Anim. Ecol. 80, 1278–1287.

Levins, R., 1963. Theory of fitness in a heterogeneous environment. II. Developmental flexibility and niche selection. Am. Nat. 97, 75–90.

Levins, R., 1968. Evolution in Changing Environments. Princeton University Press, Princeton, NJ, 132 pp.

Long, J.D., Trussell, G.C., 2007. Geographic variation in seaweed induced responses to herbivory. Mar. Ecol. Prog. Ser. 333, 75–80.

Luttbeg, B., Sih, A., 2010. Risk, resources and state-dependent adaptive behavioural syndromes. Phil. Trans. Soc. B 365, 3977–3990.

Miner, B.G., Sultan, S.E., Morgan, S.G., Padilla, D.K., Relyea, R.A., 2005. Ecological consequences of phenotypic plasticity. Trends Ecol. Evol. 20, 685–692.

Miner, B.E., De Meester, L., Pfrender, M.E., Lampert, W., Hariston, N.G., 2012. Linking genes to communities and ecosystems: Daphnia as an ecogenomic model. Proc. R. Soc. B 279, 1873.

Moran, N.A., 1992. The evolutionary maintenance of alternative phenotypes. Am. Nat. 139, 971–989.

Mougi, A., 2012. Unusual predator-prey dynamics under reciprocal phenotypic plasticity. J. Theor. Biol. 305, 96–102.

Mougi, A., Iwasa, Y., 2011. Unique coevolutionary dynamics in a predator-prey system. J. Theor. Biol. 277, 83–89.

Mougi, A., Kishida, O., 2009. Reciprocal phenotypic plasticity can lead to stable predator-prey interaction. J. Anim. Ecol. 78, 1172–1181.

Mougi, A., Kishida, O., Iwasa, Y., 2011. Coevolution of phenotypic plasticity in predator and prey: why are inducible offenses rarer than inducible defenses? Evolution 65, 1079–1087.

Mykles, D.L., Ghalambor, C.K., Stillman, J.H., Tomanek, L., 2010. Grand challenges in comparative physiology: integration across disciplines and across levels of biological organization. Integr. Comp. Biol. 50, 6–16.

Neo, M.L., Todd, P.A., 2011. Predator-induced changes in fluted giant clam (Tridacna squamosa) shell morphology. J. Exp. Mar. Biol. Ecol. 397, 21–26.

Padilla, D.K., 2001. Food and environmental cues trigger an inducible offence. Evol. Ecol. Res. 3, 15–25.

Padilla, D.K., Adolph, S.C., 1996. Plastic inducible morphologies are not always adaptive: the importance of time delays in a stochastic environment. Evol. Ecol. 10, 105–117.

Paul, V.J., Van Alstyne, K.L., 1992. Activation of chemical defenses in the tropical green-algae Halimeda spp. J. Exp. Mar. Biol. Ecol. 160, 191–203.

Pavia, H., Brock, E., 2000. Extrinsic factors influencing phlorotannin production in the brown alga Ascophyllum nodosum. Mar. Ecol. Prog. Ser. 193, 285–294.

Pavia, H., Toth, G.B., 2000. Inducible chemical resistance to herbivory in the brown seaweed Ascophyllum nodosum. Ecology 81, 3212–3225.

Pavia, H., Cervin, G., Lindgren, A., Aberg, P., 1997. Effects of UV-B radiation and simulated herbivory on phlorotannins in the brown alga Ascophyllum nodosum. Mar. Ecol. Prog. Ser. 157, 139–146.

Peacor, S.D., Pangle, K.L., Schiesari, L., Werner, E.E., 2012. Scaling-up anti-predator phenotypic responses of prey: impacts over multiple generations in a complex aquatic community. Proc. R. Soc. B 279, 122–128.

Peckol, P., Krane, J.M., Yates, J.L., 1996. Interactive effects of inducible defense and resource availability on phlorotannins in the North Atlantic brown alga Fucus vesiculosus. Mar. Ecol. Prog. Ser. 138, 209–217.

Piersma, T., Drent, J., 2003. Phenotypic flexibility and the evolution of organismal design. Trends Ecol. Evol. 18, 228–233.

Pohnert, G., Steinke, M., Tollrian, R., 2007. Chemical cues, defence metabolites and the shaping of pelagic interspecific interactions. Trends Ecol. Evol. 22, 198–204.

Pullin, A.S., Stewart, G.B., 2006. Guidelines for systematic review in conservation and environmental management. Conserv. Biol. 20, 1647–1656.

Reed, T.E., Waples, R.S., Schindler, D.E., Hard, J.J., Kinnison, M.T., 2010. Phenotypic plasticity and population viability: the importance of environmental predictability. Proc. R. Soc. B 277, 3391–3400.

Relyea, R.A., 2002. Costs of phenotypic plasticity. Am. Nat. 159, 272–282.

Rohde, S., Wahl, M., 2008. Temporal dynamics of induced resistance in a marine macroalga: time lag of induction and reduction in Fucus vesiculosus. J. Exp. Mar. Biol. Ecol. 367, 227–229.

Rohde, S., Molis, M., Wahl, M., 2004. Regulation of anti-herbivore defence by Fucus vesiculosus in response to various cues. J. Ecol. 92, 1011–1018.

Santelices, B., Varela, D., 1993. Intra-clonal variation in the red seaweed Gracilaria chilensis. Mar. Biol. 116, 543–552.

Scheiner, S.M., 1993. Genetics and evolution of phenotypic plasticity. Annu. Rev. Ecol. Syst. 24, 35–68.

Schwenk, K., Padilla, D.K., Bakken, G.S., Full, R.J., 2009. Grand challenges in organismal biology. Integr. Comp. Biol. 49, 7–14.

Sih, A., Bolnick, D.I., Luttbeg, B., Orrock, J.L., Peacor, S.D., Pintor, L.M., Preisser, E., Rehage, J.S., Vonesh, J.R., 2010. Predator-prey naivete, antipredator behavior, and the ecology of predator invasions. Oikos 119, 610–621.

Smith, L.D., Palmer, A.R., 1994. Effects of manipulated diet on size and performance of brachyuran crab claws. Science 264, 710–712.

Svensson, C.J., Pavia, H., Toth, G.B., 2007. Do plant density, nutrient availability, and herbivore grazing interact to affect phlorotannin plasticity in the brown seaweed Ascophyllum nodosum. Mar. Biol. 151, 2177–2181.

Tollrian, R., 1993. Neck teeth formation in Daphnia pulex as an example of continuous phenotypic plasticity: morphological effects of Chaoborus kairomone concentration and their quantification. J. Plankton. Res. 15, 1309–1318.

Tollrian, R., 1995. Predator-induced morphological defenses—costs, life-history shifts, and maternal effects in *Daphnia pulex*. Ecology 76, 1691–1705.

Tollrian, R., Dodson, S.I., 1999. Inducible defenses in cladocerans. In: Tollrian, R., Harvell, C.D. (Eds.), The Ecology and Evolution of Inducible Defenses. Princeton University Press, Princeton, NJ, pp. 177–202.

Tollrian, R., Harvell, C.D., 1999. The Ecology and Evolution of Inducible Defenses. Princeton University Press, Princeton, NJ, 395 pp.

Toth, G.B., Pavia, H., 2007. Induced herbivore resistance in seaweeds: a meta-analysis. J. Ecol. 95, 425–434.

Vermeij, G.J., 1978. Biogeography and Adaptation. Patterns of Marine Life. Harvard University Press, Cambridge, MA, 332 pp.

Vermeij, G.J., 1987. Evolution and Escalation: An Ecological History of Life. Princeton University Press, Princeton, NJ, 549 pp.

Vermeij, G.J., 1994. The evolutionary interaction among species—selection, escalation, and coevolution. Annu. Rev. Ecol. 25, 219–236.

Wade, M.J., 2007. The co-evolutionary genetics of ecological communities. Nat. Rev. Genet. 8, 185–195.

Weber, M.G., Agrawal, A.A., 2012. Phylogeny, ecology, and the coupling of comparative and experimental approaches. Trends Ecol. Evol. 27, 394–403.

West-Eberhard, M.J., 1989. Phenotypic plasticity and the origins of diversity. Annu. Rev. Ecol. Syst. 20, 249–278.

West-Eberhard, M.J., 2003. Developmental Plasticity and Evolution. Oxford University Press, Oxford, UK, 794 pp.

Woltereck, R., 1909. Weitere experimentelle Untersuchungen über Artveränderung, speziell über das Wesen quantitativer Artunterschiede bei *Daphnien*. Verh. Deutsch. Zool. Ges. 19, 110–173.

CHAPTER THREE

The Biology, Ecology and Fishery of the Dungeness crab, *Cancer magister*

Leif K. Rasmuson[1]

University of Oregon, Oregon Institute of Marine Biology, P.O. BOX 5389, Charleston, Oregon, USA
[1]Corresponding author: e-mail address: Rasmuson@uoregon.edu

Contents

Abstract

The Dungeness crab, *Cancer magister*, is a commercially important crustacean that ranges from the Pribilof Islands, Alaska, to Santa Barbara, California. Mating occurs between recently moulted females and post-moult males. After approximately 90 days, females release planktonic larvae into the water column. Stage-I zoea are found in the nearshore environment and subsequent zoeal stages are found at greater distances. After approximately 80 days, zoea moult into megalopa, which move first from the open ocean onto the continental shelf and then across the shelf to settle in the nearshore environment or estuaries. Crabs reach sexual maturity at 2–3 years of age. The fishery for *C. magister* is managed using a 3-S management strategy which regulates catch based on size, sex and season. As more fisheries seek sustainability certifications, the

Advances in Marine Biology, Volume 65
ISSN 0065-2881
http://dx.doi.org/10.1016/B978-0-12-410498-3.00003-3
95

Dungeness crab fishery presents an excellent test case of how to sustainably manage a crustacean fishery.

Keywords: Dungeness crab, *Cancer magister*, *Metacarcinus magister*, Crab fishery, Decapod, Crustacean, Invertebrate fishery

1. INTRODUCTION

The Dungeness crab was originally described in 1852 by James Dana (Dana, 1852; Jensen and Armstrong, 1987). Recent publications (Wicksten, 2009) have referred to the Dungeness crab as *Metacarcinus magister*, based on morphological studies by Schweitzer and Feldmann (2000), who elevated the older subgenus name *Metacarcinus*, established in 1862, to full genus level. However, molecular work by Harrison and Crespi (1999) does not support the monophyly of the genus *Metacarcinus* nor of some other cancrid genera used by Schweitzer and Feldmann (2000). Thus, I follow Kuris et al. (2007) in retaining the name *Cancer magister*. Both Kuris et al. (2007) and Wicksten (2009) provide excellent dichotomous keys to the adult cancrid crabs of the California and Alaska Currents. As with all brachyurans, males have a narrow pointed abdomen, while females have a broader, rounded abdomen.

This chapter updates historic reviews and synthesizes the literature on the biology, ecology and fishery of *C. magister* throughout its range (Fisher and Velasquez, 2008; Melteff, 1985; Pauley et al., 1986; Wild, 1983c). Since *C. magister* ranges from the Pribolof Islands, Alaska, to Santa Barbara, California, the organism inhabits both open oceans (California and Alaska Currents) and inland fjords (Puget Sound through the inside passage) (Figure 3.1; Jensen and Armstrong, 1987). Due to the multitude of environments that *C. magister* lives in, the biology of the organism changes over the species range (Table 3.1). For example, off California and Oregon, larvae are released during the winter, while in Alaska, larvae are released during the summer (Jaffe et al., 1987). Therefore, throughout this review, I state where research was conducted, so readers can view the studies in the broad context of the entire population. Historic reviews have focused on the population in the California Current and relatively little coverage of the Alaska Current has been given. Many of these reviews have primarily focused on the fishery and only briefly touched on the biology and ecology of *C. magister*. Thus, in this review, I first provide a brief introduction to the habitats *C. magister* occupies, followed by an extensive review of the biology and ecology of the species. I conclude by reviewing the management of the fishery, the impacts of the fishery (both direct and indirect) and recent breakthroughs in catch prediction.

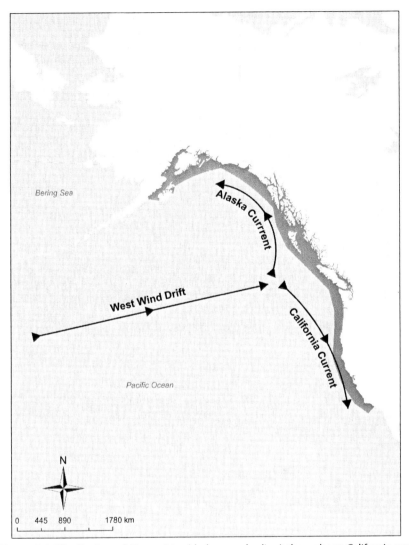

Figure 3.1 Distribution of *C. magister* (dark grey shading) throughout California and Alaska Currents. Arrows denote the general direction of currents. *Base map provided by ESRI (2011).*

2. HABITAT AND OCEANOGRAPHY

The Alaska and California Currents originate where the West Wind Drift collides with Vancouver Island and divides into currents that move north as the Alaska Current and south as the California Current (Figure 3.1; Hickey,

Table 3.1 Peak reproductive timing throughout the range of *C. magister*

Location	Moulting/ mating	Egg deposition	Hatching	Larval duration (range of time)	Settlement
California	March–June	September–November	December–February	115 (105–125)	April–May
Oregon–Washington	March–June	October–December	January–March	130 (89–143)	April–August
Puget Sound	April–September	October–December	February–May	150	June–August
British Columbia	No data	September–February	December–June	No data	July–Later
Alaska	June–July	September–November	April–August	154 (146–162)	September–October

See text for citations

1979; Mann and Lazier, 2006). The location of the bifurcation shifts to the
south (~47 °N) in the winter and northwards to ~50 °N in the summer.
Both the California and Alaska Currents are subject to large inter- and
intra-annual variations that affect the hydrodynamics of the ecosystem
(Huyer et al., 1979; Mantua and Hare, 2002). A large multiyear cycle known
as the Pacific Decadal Oscillation (PDO) is driven by changes in the amount
of water transferred into the California and Alaska Currents from the West
Wind Drift (Minobe and Mantua, 1999). During a cool (negative) phase
PDO—characterized by colder than average water temperatures in the
Northeast Pacific ocean—more cold water is shifted into the California
Current and southward flow is enhanced. During a warm (positive) phase,
the converse is true and more water is shifted into the Alaska Current and
southward flow in the California Current is decreased. PDO phase affects
all trophic levels throughout the California and Alaska Current ecosystems
(Hooff and Peterson, 2006; Keister et al., 2011; Mantua et al., 1997).

2.1. California Current

2.1.1 Oceanography

The California Current is the eastern boundary of the North Pacific Subtropical Gyre and is characteristically broad (~500 km wide) and slow-
moving (5–10 cm s^{-1}) (Bakun, 1996; Strub and James, 1988). Below the
California Current on the continental slope is a poleward counter current
known as the California Undercurrent (Hickey, 1979). The undercurrent

flows at depths of 200–300 m with a mean velocity of 10 cm s^{-1} (Collins et al., 2000; Pierce et al., 2000; Reed and Halpern, 1976). On an intra-annual level, changes in atmospheric pressure systems cause seasonal changes in winds and currents (Lynn and Simpson, 1987). During the winter, winds blow towards the north, creating an oceanic surface current (known as the Davidson Current or Inshore Counter Current) that flows north at a mean velocity of 15 cm s^{-1} (Austin and Barth, 2002; Huyer et al., 1989; Strub and James, 1988). While the Davidson Current is flowing northwards, the California Current is still flowing southward off the continental shelf. Our understanding of how far off the shelf northwards flow occurs is minimal; northwards flow is reported as far as 300 km from shore in some areas and in others to be restricted to the continental shelf (Hickey, 1979). Over the course of about 1 week in the spring, during an event known as the spring transition, winds start blowing towards the south and the California Current begins flowing south over the continental shelf (Huyer, 1977). The California Current ecosystem is a monsoonal upwelling system driven by the change in the location of the North Pacific High (Huyer, 1983). Following the spring transition, during the spring and summer, winds are characteristically upwelling favourable, while during the fall and winter, winds are downwelling favourable.

2.1.2 Habitat

In the California Current, unconsolidated sediments (sand, mud and sand/mud mixtures) are 4.5 times more abundant than hard substrates (Romsos, 2004). On the continental shelf, soft sediments (the habitat of C. *magister*) account for ~53% of the bottom, the majority of which is sandy. Although C. *magister* often preferentially settle in estuaries, there are relatively few estuaries and inlets in the California Current system, and thus, the majority of C. *magister* reside in the open ocean.

2.2. Alaska Current

2.2.1 Oceanography

Where the West Wind drift collides with Vancouver Island, the ocean is characterized by 'confused' (non-directional) currents with numerous eddies (10–25 cm s^{-1}) and meanders (Thomson, 1981). Along Vancouver Island, a summertime nearshore current flows northwest along the continental shelf. The Alaska Current flows northwards off the shelf at an average velocity of 25 cm s^{-1} (Thomson, 1981). The same change in atmospheric pressure that influences the California Current causes the Alaska Current to accelerate northwards during the winter. The interaction between freshwater runoff from the numerous rivers in Alaska and British Columbia and winds causes the circulation of the Alaska Current to be variable (Stabeno et al., 2004). The Alaska Current is characteristically downwelling favourable with mean velocities of $\sim$30 cm s^{-1} (Favorite, 1967). The current flows northwards into the Gulf of Alaska where it turns eastward as the strong Alaska Stream. The Alaska Stream then flows along the Aleutian Peninsula until it collides with the Oyashio Current and flows southward. Extending from Puget Sound (Washington) northwards through the Gulf of Alaska is a network of inland waters with complex circulation driven by tidal currents and freshwater input. The complex circulation patterns likely have significant effects on the biology of C. magister especially as the larvae disperse.

2.2.2 Habitat
I was unable to find any literature on the subtidal habitats of the Alaska Current. A recent publication by the National Marine Fisheries Service's Alaska Fisheries Science Center has discussed future plans to fill this gap in knowledge (Sigler et al., 2012).

3. BIOLOGY AND ECOLOGY
3.1. Reproduction
3.1.1 Mating
In brachyurans, mating occurs between recently moulted (soft) females and males that have already moulted and since re-hardened (Figure 3.2; Hartnoll, 1969; Snow and Nielson, 1966). Dungeness crabs reach sexual maturity at a carapace width of 100 mm, which occurs at 2 years of age in Humboldt County, California; however, in Alaska, crab gonads are not fully developed at a carapace width of 100 mm, so eggs are not extruded until the following year ($\sim$3 years old) (Butler, 1961; Cleaver, 1949; Hankin et al., 1989; Scheding et al., 1999). When females are close to moulting, males become more active and move towards the nearshore (Barry, 1985; Cleaver, 1949; MacKay, 1942). Males will grasp and carry a female that is close to moulting for up to 2 weeks in a 'premating embrace'. The data conflict about what happens when the female is ready to moult: either (1) the female is released by the male and moults outside the embrace of the male (Cleaver, 1949) or (2) the female moults while being embraced by the male (Butler, 1960; Cleaver, 1949; Snow and Nielson, 1966). After the female

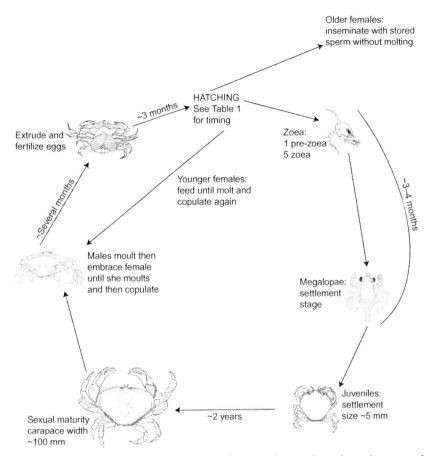

Figure 3.2 Life cycle of *C. magister*. Timing is relative and varies throughout the range of the species (see Table 3.2 for timings).

moults, the male stands over the female who lies with her dorsal side on the substrate and they both extend their abdomens out and away from their bodies. Then using modified pleopods, the male deposits spermatophores into the female gonopores. Following copulation, some of the seminal fluid hardens into a sperm plug which prevents other males from mating with the female (Jensen et al., 1996). Since males mate multiple times each year and the sex ratio of males to females is 1:1, competition for females is high and thus the sperm plug reduces sperm competition (Oh and Hankin, 2004; Orensanz and Gallucci, 1988). Females who did not undergo an annual moult were found to have retained sperm and researchers hypothesized that the sperm remain viable for up to 2.5 years (Hankin et al., 1989) and recent molecular analysis corroborates this hypothesis (Jensen and Bentzen, 2012).

Researchers have speculated that intensive fishing on male crabs reduces reproductive output of females (Smith and Jamieson, 1991b). To test this, two techniques have been suggested to determine whether or not females mated during the previous mating season. First, the process of embracing during mating injures the limbs of the crabs which cause scarring, and early work suggested that the presence of mating scars could be used to assess mating success (Cleaver, 1949). A recent in-depth examination of this technique determined that not all crabs are scarred during mating and that limbs are also scarred by other processes, making the technique ineffective (Ainsworth, 2006). The second technique is more promising. Sperm plugs are found to still be present in the female reproductive tract 180 days after copulation, and thus, dissecting females to look for sperm plugs provides a reliable indicator of whether or not females mated that year (Oh and Hankin, 2004).

Studies in Northern California and Oregon that used sperm plugs to indicate mating success determined that reproductive output is not limited and ~83% of females showed evidence of successful mating the previous year (Dunn and Shanks, 2012; Hankin et al., 1997). Mating success in Hood Canal, Washington, an enclosed glacial fjord, was comparable (~80% of females had mated) to values in the open ocean (L.K. Rasmuson, unpublished work).

3.1.2 Brooding

In Central and Northern California, the weight of ovarian tissue increases linearly until the eggs are extruded (Wild, 1983a). A few months following copulation, females extrude their eggs, thereby inseminating them, and attach them to the setae on the pleopods below the abdominal flap (Figure 3.2; Wild, 1983b). When the eggs are extruded, they are bright orange (Buchanan and Milleman, 1969). The timing of egg deposition varies greatly throughout *C. magister's* range, with females becoming ovigerous from September–November in California, October–December in Oregon and Washington, September–February in British Columbia and September–November in Alaska (Table 3.1; Jaffe et al., 1987; Shirley et al., 1987). Laboratory studies in Alaska have demonstrated that female crabs do not extrude eggs annually (Swiney and Shirley, 2001). In a comprehensive follow-up to their 2001 work, Swiney et al. (2003) conducted a field study that corroborated their laboratory finding that females do not extrude their eggs annually. They suggest that there are two reproductive pathways based on the size of female crabs: (1) large (carapace width >141 mm) females store sperm but do not moult or extrude eggs annually and (2) smaller females (carapace width <141 mm) moult, mate and extrude eggs annually. Larger female crabs extrude their eggs later in the season than smaller crabs and thus their eggs hatch later in the season. The most probable explanation is that the time period between hatching and mating is too short for the gonads of large females (carapace width >141 mm) to fully develop, forcing them to spawn every other year (Table 3.1). Whereas the earlier hatching times of smaller females provide a sufficient time period between hatching and moulting for their gonads to fully develop.

Females carry between 1.5 and 2.5 million eggs and there is no correlation between carapace width and fecundity (Hankin et al., 1989; Wickham, 1979b,c). Due to the large number of eggs that are extruded, the abdominal flap is raised significantly from the cephalothorax limiting female movement (L.K. Rasmuson, personal observation). Females must bury themselves in sand 5 to 10 cm deep in order for eggs to remain attached to the setae (Fisher and Wickham, 1976; Wild, 1983b). After egg extrusion, females in enclosed waters (Puget Sound and Alaska) migrate to shallow water and form dense aggregations (Armstrong et al., 1988; Scheding et al., 1999). In Puget Sound, aggregations of brooding females were observed in 1 to 3 m depth of water within dense eelgrass bands, whereas in Alaska, dense aggregations of brooding females occur at a depth of ~16 m (Armstrong et al., 1988; Stone and O'Clair, 2002). Anecdotes from the Washington coast during winter razor clam (*Siliqua patula*) openings describe numerous ovigerous females in the surf zone (Northrup, 1975). It is likely that females in the coastal ocean form aggregations, but the depth at which this occurs is unknown (Diamond and Hankin, 1985). A 12-year time series of brooding location collected in an enclosed fjord in Alaska, determined that females returned to a specific site characterized by unconsolidated, homogeneous, highly permeable sand (Stone and O'Clair, 2002).

Wild (1983b) found a negative linear correlation between water temperature and the length of time that egg masses were brooded. Brooding times ranged from 130 days at 9 °C to 65 days at 17 °C. Although rate of development decreased with a decrease in temperature, the researcher noted that as temperature rose from 13 to 17 °C, fewer eggs were produced and hatching success declined. Mayer (1973) suggested that at a salinity of 25, 12 °C may represent the maximum temperature at which eggs develop normally.

3.2. Larval biology

3.2.1 Hatching

Hatching occurs when prezoea are fully developed, at which time the egg masses are dark brown (Buchanan and Milleman, 1969). Timing of hatching varies over the range of *C. magister* and occurs in late December–February in Central California, January–March in Northern California and Oregon, February–May in Puget Sound, December–June on the outer coast of British Columbia, April in the Queen Charlotte Islands and April–August in Alaska (Table 3.1; Fisher, 2006; Jaffe et al., 1987; Shirley et al., 1987; Swiney and Shirley, 2001). Many species of crabs synchronize the release of their larvae to specific tides and light levels; however, no study has yet examined the larval release patterns of *C. magister* (Morgan, 1995; Stevens, 2003, 2006). Plots of hatching date from laboratory studies suggest that there are no endogenous cues to larval release (Wild, 1983b). *In situ* studies should be conducted to ascertain whether or not larval release is synchronized.

The biology and dispersal of the larval stages of *C. magister* are well studied. The larvae are often released as prezoea, and initially, researchers thought prezoea released into the water column did not survive, but more recent studies have demonstrated that prezoea survive and develop into stage-I zoea (Buchanan and Milleman, 1969; MacKay, 1942; Poole, 1966). In the laboratory, the transition from prezoea to stage-I zoea takes only a few seconds, and thus prezoea are only in the water column for seconds to minutes. *C. magister* then progress through five zoeal stages and one megalopal stage in the water column (Figure 3.2; Poole, 1966). All five zoeal stages of *C. magister* have large compound eyes, four spines (one dorsal, one rostral and two lateral) and swim by flexing their maxillipeds. Zoeal spine lengths increase as incubation temperature decreases (Shirley et al., 1987). The megalopa of *C. magister* has large compound eyes, two spines (one dorsal and one rostral) and swim with their pleopods. A set of intermoult stages have been developed for megalopae that allow researchers to determine the relative age of megalopae (Hatfield, 1983).

In a multiyear project, the Oregon Fish Commission worked to develop methods for culturing the larvae of *C. magister* (Gaumer, 1969, 1970, 1971; Reed, 1966, 1969). While rearing larvae, the Oregon Fish Commission tested the effects of multiple factors on the growth and development of zoea. The development of zoea is influenced by water temperature and to a lesser extent salinity. Normal development occurs over a temperature range of 10.0–13.9 °C and a salinity range of 25–30. The duration of each larval stage decreases as water temperature increases, though at temperatures >14 °C, mortality increases (Ebert et al., 1983). *C. magister* from Oregon were used for these studies, so the results likely apply to the open ocean population in the California Current and not to populations in inland waters and/or the Alaska Current.

It appears that although temperatures within the range of 14–22 °C do not affect the number of juveniles that metamorphose into adults (Sulkin et al., 1996), temperature may influence larval survival in the open ocean. Sulkin and McKeen (1989) used crabs from Puget Sound to examine the potential effects of elevated water temperatures on development. They tested higher water temperatures (10, 15 and 20 °C) and determined that survival was highest for zoea reared in 10 °C water, but the duration of each stage drastically decreased as water temperature increased (Table 3.2). Sulkin and McKeen (1996) examined historic temperature records and mimicked weekly temperatures from the open ocean (~10–12 °C) and Puget Sound (~7–12 °C) in the laboratory. They determined that zoea reared at Puget Sound water temperatures were in the water column 44% longer than larvae raised at open ocean temperatures.

Table 3.2 Effect of temperature on the day of moulting (mean day), length of the stage (days) and percentage of larvae surviving

Stage	Temperature	Mean day	Length of stage	Percent of population surviving
Zoea I	10	13.2	13.2	87
	15	8.3	8.3	85
	20	7.5	7.6	72
Zoea II	10	24.5	11.3	83
	15	14.3	6.2	82
	20	13	5.5	62
Zoea III	10	37.1	12.7	79
	15	20.8	6.8	75
	20	18.8	6.1	57
Zoea IV	10	50.8	13.7	71
	15	28.2	7.3	66
	20	25.1	6.9	44
Zoea V	10	68.9	18.8	46
	15	38.5	10.4	27
	20	NA	NA	0

Adapted from Sulkin and McKeen (1989).

Moloney et al. (1994) combined results from multiple laboratory studies on the development rate and mortality of larvae at different temperatures and salinities to generate a numerical model of development that was combined with historically accurate simulations of water temperatures and salinities in the California Current. Modelled larval duration ranged from 74 to 163 days depending on latitude, which is slightly different than the measured durations (Table 3.1). Additionally, they argue that it is inaccurate to assume that mortality of larvae within the plankton is constant, and additional work is needed to determine what the mortality rate of larvae is while they are in the plankton. Their results demonstrate that the influence of water temperature and salinity can alter the rate of larval development by a factor of two. The extended length of development in colder water may increase the overall mortality of larvae and thus may explain inter-annual variation and north–south variation in the population. They note that their work only applies to open ocean populations in the California Current and that enclosed populations in areas such as Puget Sound are not represented.

3.2.2 Diet

Laboratory studies have found that unfed larvae and larvae fed only in the first 24 h after hatching can subsist on their yolk reserves for approximately 15 days before they die (Reed, 1969; Sulkin et al., 1998a). Attempts to rear zoea on a diet solely of diatoms were unsuccessful (Hartman and Letterman, 1978). Zoea that were fed mussel larvae (*Mytilus edulis*) did not survive but did well when fed barnacle larvae (*Balanus glandula*) (Gaumer, 1971; Reed, 1969). Zoea fed diets of brine shrimp (*Artemia* sp.) successfully metamorphosed, and when diatoms (*Skeletonema* sp.) were added to the diet, survival was >88%; however, if brine shrimp concentration exceeded 5 shrimp ml^{-1}, then survival of zoea decreased (Gaumer, 1971; Hartman, 1977). The larvae of *C. magister* are capable of feeding in the dark, suggesting they are not ocular hunters (Sulkin et al., 1998a). A short period of feeding each 24-h period is sufficient in preventing mortality of larvae (Sulkin et al., 1998a). Early zoea of *C. magister* feed on protists that naturally occur in the water column (Sulkin et al., 1998b). Stable isotope work on wild megalopae suggests that they are omnivorous, which coincides with findings from the aforementioned laboratory studies (Kline, 2002). Larvae also commonly consume heterotrophic prey that ingest toxic algae, and thus, researchers examined the effect of the toxins on the survival of zoea (Garcia et al., 2011). Results indicated that fewer zoea survived that consumed toxic prey than those that did not, and not surprisingly individuals that had consumed toxic prey and survived remained at each stage longer than those that did not. Following up, researchers have demonstrated that the consumption of toxic food sources decreases larval survival not because the food is toxic, but rather the food is nutritionally deficient (Burgess, 2011).

3.2.3 Predation

No literature reports selective feeding by pelagic invertebrates or fishes on the zoea of *C. magister*. In the field, I have dissected English sole (*Parophrys vetulus*) and found their stomachs completely full of stage-I *C. magister* larvae (L.K. Rasmuson, personal observation). The megalopae of *C. magister* are consumed by a variety of fish species such as coho salmon (*Oncorhynchus kisutch*), chinook salmon (*Oncorhynchus tshawytscha*) and hake (*Merluccius productus*) (Botsford et al., 1982; Emmett and Krutzikowsky, 2008; Methot, 1989).

3.2.4 Larval behaviour and swimming

The zoea of *C. magister* respond to light in laboratory studies, moving deeper in the water column as light intensity increases (Gaumer, 1971; Jacoby, 1982). In the laboratory, zoea swim into currents (positive rheotaxis), and megalopae have slightly stronger rheotaxis than zoea (Gaumer, 1971). Ninety-five percent of megalopae observed *in situ* displayed strong positive rheotaxis (L.K. Rasmuson, in preparation). As megalopae approach settlement, they are attracted to light (positive phototaxis) and cling to objects they encounter while swimming (thigmokinesis) (Hatfield, 1983; Reilly, 1983a).

Reported values of swimming speed for different larval stages vary between studies; however, all studies demonstrate that compared to many other planktonic organisms, *C. magister* are strong swimmers (Fernandez et al., 1994b; Gaumer, 1971; Jacoby, 1982). In general, early zoea (I–III) are capable of swimming at speeds ranging from 0.58 to 0.95 cm s^{-1} (Gaumer, 1971; Jacoby, 1982), while later-stage zoea (IV and V) are capable of swimming at a speed of 1.5 cm s^{-1}. *In situ* swimming speeds of the megalopae average 12 cm s^{-1} with a range of 5–20 cm s^{-1} (L.K. Rasmuson, in preparation), but swimming speeds of megalopae determined in the laboratory are more variable and range from 4.2 to 44 cm s^{-1} (Fernandez et al., 1994b; Jacoby, 1982).

3.2.5 Vertical migration

Zoea and megalopae vertically migrate and occupy the neuston at night and/or early evening, returning to deeper waters during the day (Booth et al., 1986). The depth they occupy during the day has eluded researchers for many years. In a comparative study of the Puget Sound (inland waters) and the open ocean off Vancouver Island (on the continental shelf), researchers determined that late intermoult stage megalopae were migrating to depths of ~160 m in Puget Sound and ~25 m in the open ocean during the day (Jamieson and Phillips, 1993). Off the continental shelf, megalopae migrate to depths >70 m, and upon returning to the continental shelf, may stop vertically migrating (A.L. Shanks and G.C. Roegner, unpublished data). Puget Sound megalopae are smaller and settle later than oceanic megalopae, and researchers speculate that differences in vertical migration depth between oceanic and inland megalopae may aide in the retention of larvae within Puget Sound which could be the cause of the overall small size of Puget Sound megalopae. (Hobbs and Botsford, 1992; Lough, 1976; Reilly, 1983a). In Alaska, *C. magister* zoea and megalopae may undergo a crepuscular (occupying the surface only at dusk/dawn) rather than diel migration (Park and Shirley, 2005).

3.2.6 Cross-shelf distribution

The movement of *C. magister* larvae within enclosed waters (e.g. Puget Sound, British Columbia and Alaska Fjords) and the Alaska Current has not been studied enough to provide a description of the movement of larvae. Thus, this section pertains to the open ocean population in the California Current. Stage-I larvae are released within 8 km of shore and migrate off the continental shelf as they develop (Lough, 1976; Reilly, 1983a). Stages I and II zoea are commonly found on the continental shelf, while stages III–V are found off the continental shelf at distances >150 km (Reilly, 1983a). While over the continental shelf, larvae will be transported northwards by the Davidson Current, but as they migrate off the shelf, they may enter the California Current (depending on the distance of southward flow from shore) and be transported southward. The majority of stage V larvae are concentrated 50–100 km from shore (Reilly, 1983a). After migrating off the shelf, the zoea moult into megalopae. Although megalopae can be found at great distances offshore, they must settle in the nearshore habitat (Jamieson and Phillips, 1988). Thus, there are mechanisms that advect the megalopae of *C. magister* from seaward of the continental shelf to shelf waters and subsequently back to the nearshore environment.

3.2.7 Dispersal patterns

The commercial catch of *C. magister* on the outer coast of California, Oregon and Washington has undergone many oscillations, and considerable research has attempted to explain them (Methot, 1989). I will discuss later why these oscillations are not likely caused by intensive fishing pressure. Many researchers speculate that the effects of oceanographic conditions on dispersal and recruitment of larvae are the cause of these fluctuations. Thus, many studies have examined the influence of oceanography in the California Current by correlating commercial catch to physical factors (Table 3.3). Since it takes approximately 4 years for a Dungeness crab to recruit from a larva to the fishery, many researchers correlate commercial catch with physical indices that occurred 4 years prior to the fishery (Hackett et al., 2003). These numerous studies provide us with an idea of how the larvae of *C. magister* disperse in the ocean.

In his thesis, Lough (1975) enumerated plankton from samples off Newport Oregon over a 2-year period. He determined that initially, larvae are released into the Davidson Current and swept north until the spring transition occurs and at which point larvae are swept south with the California Current. However, as discussed earlier, we are unsure how far off the shelf larvae migrate and how far off the continental shelf the Davidson Current occurs, so it is possible that the cross-shelf migration of larvae moves them from the Davidson Current into the California Current before the spring transition occurs shifting the direction larvae are advected. Upwelling indices are correlated to commercial catch with a time lag of 0.5–1.5 years (Peterson, 1973). However, since it takes 4 years for *C. magister* to recruit to the fishery, this correlation does not demonstrate any effects of upwelling on the dispersal of larvae (Botsford and Wickham, 1975; Peterson, 1973). Correlations between catch and sunspot number have also been attempted, though other articles suggest that this publication may have been in jest and catch patterns cannot be explained by sunspots (Hankin, 1985; Love and Westphal, 1981). A study off California suggests that stage-I zoea are transported offshore by a combination of estuarine runoff and upwelling circulation, though there are relatively few estuaries in the California Current, and most upwelling occurs a few months after larvae are released, implying that these hypotheses are likely incorrect (Reilly, 1983a). Comparing catch to wind data, researchers found a correlation (with a 4-year lag) with southward wind stress (Johnson et al., 1986). Recalculating the wind stress reported in Johnson et al. (1986) and including data on the distribution (including vertical migration) of megalopae, megalopal abundance in the nearshore environment was shown to be correlated with onshore winds (Hobbs et al., 1992). Off British Columbia (just north of the California Current), for multiple years there were no recruitment events of *C. magister* even

Table 3.3 Influence of physical and biological factors on the biology of *C. magister*

Year	Findings
Physical factors	
Peterson (1973)	Time lag 0.5–1.5 years between upwelling index and catch (does not correlate with recruitment). Suggest the correlations due to food availability
Botsford and Wickham (1975)	Time lag 9–12 years does not correlate with recruitment; suggest correlation due to food availability
Love and Westphal (1981)	Correlation to sunspot number; other researchers suggest this chapter may have been written in jest
Wild (1983b)	Low catch years correlated with warmer winter water temperatures (and weaker southward transport) 4 years earlier
Johnson et al. (1986)	Correlation between catch and southward wind 4 years earlier
McConnaughey et al. (1992)	Increased megalopae in nearshore environment during years of increased shoreward transport
Hobbs et al. (1992)	Increased settlement of larvae during years with decreased northwards transport
McConnaughey et al. (1994)	Suggest that larvae may be retained in the nearshore environment rather than being transported offshore
Botsford and Lawrence (2002)	Commercial catch correlated with overall cooler conditions in the California Current
Shanks and Roegner (2007)	Recruitment and commercial catch correlated with day of the year of the spring transition and amount of returning larvae
Shanks et al. (2010)	Recruitment and commercial catch negatively correlated with Pacific Decadal Oscillation index; population limited by recruitment at beginning until levels off and possibly cannibalism affects recruitment
Shanks (in press)	Recruitment and commercial catch correlated with amount of upwelling following spring transition. Catch did indeed level off when recruitment was approximately 1 million megalopae in 1 year
Biological factors	
Botsford and Wickham (1978)	Suggested that fluctuations may be caused cannibalism. Botsford (1984) stated if cannibalism was the cause the cycles would likely have been more stable

Table 3.3 Influence of physical and biological factors on the biology of *C. magister*—cont'd

Year	Findings
Wickham (1979b)	Suggested predation by *C. errans* may impact reproductive output, but the effect is not big enough to cause fluctuations
McKelvey et al. (1980)	Density-dependent egg/larval survival influenced by production
Botsford et al. (1982)	No correlation between catch and Chinook and Coho Salmon catch 4 years earlier
Botsford et al. (1983)	No correlation between catch by humans and fluctuations in population
(Shanks, in press)	Suggested that when recruitment is high, cannibalism and competition among juvenile crabs impact the population

though many megalopae were found in the neuston on the continental shelf which caused researchers to suggest that the Vancouver Coastal Current flowing northwards in the opposite direction of the shelf break current acts as a barrier to the transport of megalopae across the continental shelf (Jamieson et al., 1989). Thus, megalopae only make it back to the nearshore environment when the Vancouver Coastal Current relaxes. Additionally, they examined surface drifter tracks and noted that the drifters were transported across the shelf by winds from the south and thus hypothesize that megalopae may be transported across the shelf by similar winds.

Over a 5-year period, McConnaughey et al. (1992) used a modified beam trawl in estuaries and on the continental shelf to collect recently settled juveniles. They correlated recruit density to oceanic indices and concluded that westward Ekman transport may not be transporting zoea off the shelf. Additionally, they found a negative correlation between the number of settlers and the amount of northward alongshore transport. They demonstrated that in years when larvae are transported further north by the Davidson Current, recruitment in the California Current is limited. Further they hypothesize that since larvae are initially transported northwards, it is possible that they could be transported into the Alaska Current and thus do not move southward when the Davidson Current disappears (spring transition). Thus, the geographic closeness of the Washington coast to the Alaska Current may mean that the populations in Washington are dependent on recruitment from populations that are further south. In follow-up work, they found a positive correlation with recruitment and the amount of onshore winds (McConnaughey et al., 1994, 1995). They use the findings from these three studies to suggest that larvae of *C. magister* do not undergo an ontogenetic migration off the continental shelf but rather are retained in the nearshore. The hypothesis that larvae are retained in the nearshore is based on sampling of settled juveniles; however, extensive plankton sampling efforts by other researchers (Jamieson and Phillips, 1993; Jamieson et al., 1989; Reilly, 1983a) have not corroborated that larvae are retained in the nearshore.

Researchers in Alaska reported finding late-stage zoea and megalopae in their plankton samples at the time when hatching occurs in Alaska (Park et al., 2007). Based on the stage of these larvae, it was clear that they had been released much earlier, and thus likely hatched in the California Current. Therefore, the most likely explanation for the presence of these late-stage larvae is that they were transported north by the Davidson Current into the Alaska Current. These data demonstrate that there is connectivity (at least in some years) between the populations in the California and Alaska Currents.

Using the annual return of megalopae to the shore as measured by the number of megalopae caught in a shore-based light trap in Coos Bay, Oregon, Shanks and Roegner (2007) correlated oceanographic indices to the number of returning megalopae (as measured by the light trap) and commercial catch 4 years later. They determined that larval recruitment explained ~90% of the variability in the adult population from the Washington/Oregon border to San Francisco, California. Furthermore, they found a strong positive correlation between the date of the year of the spring transition and number of recruits, which suggests that the shift in currents caused by the spring transition (from the Davidson to California) strongly influences the recruitment of *C. magister* larvae. Shanks and colleagues published follow-up work (2010) examining 4 additional years of recruitment and reported a negative correlation between recruitment and the PDO index. They suggest that enhanced southward transport in the California Current during negative PDO index years may be the cause of the increased recruitment. Additionally, there appears to be a positive correlation between recruitment and the amount of upwelling that occurs following the spring transition (Shanks, in press). By combining the three physical factors that affect larval return (date of spring transition, PDO index phase and amount of upwelling), one can observe a strong three-factor linear relationship. By splitting recruitment seasons into high (>100,000) and low (<100,000) settlement years, two parallel relationships are observed. When recruitment is correlated to the date of the spring transition, there is a negative relationship and when recruitment is correlated to the amount of upwelling following the spring transition, there is a positive relationship. Shanks proposes the following conceptual model to explain these three correlations. First, he hypothesizes that a negative PDO index increases southward transport of larvae, which increases the possibility of the larvae settling in Oregon rather than further north. He hypothesizes that larvae are transported onto the shelf with the water that is brought onto the shelf by wind-driven upwelling which would mean that an earlier spring transition would result in a longer period of time that larvae can be advected onto the continental shelf. However, the amount of upwelling following the spring transition is not consistent, and thus, during years with increased upwelling, more megalopae are advected onto the continental shelf.

For both correlations, more larvae tended to recruit during negative phase PDO index years. Recently, Shanks has determined that there is a negative relationship between the number of megalopae recruiting in August and September and the PDO index from January through July, suggesting that the enhanced southward transport during negative PDO index years may transport larvae from as far north as British Columbia to Oregon (Shanks, in press).

After being transported onto the continental shelf, most likely by upwelling, larvae must migrate back to the nearshore environment to settle in the adult habitat. Historically, researchers hypothesized that megalopae may be transported across the shelf by clinging to the pleustonic (living in the surface of the water column) hydroid *Vellela vellela*, although subsequent research has disproved this hypothesis (Reilly, 1983a; Wickham, 1979c). Alternatively, the megalopae of *C. magister* have been observed to be concentrated in surface convergences on the continental shelf (L.K. Rasmuson unpublished data; Shenker, 1988). Johnson and Shanks (2002) created a daily time series of *C. magister* recruitment to an Oregon estuary and report pulsed recruitment events suggestive of cross-shelf transport by internal tides. Building on Johnson and Shanks (2002), Roegner et al. (2007) correlated daily larval settlement with multiple environmental factors and found that megalopal abundance was strongly correlated with the spring–neap tidal cycle, but settlement did not peak on the day of the spring tide, but rather occurred a few days after the spring tide. The lag in recruitment relative to the spring tide is characteristic of cross-shelf transport by internal waves (Shanks, 2006). Therefore, the researchers' findings corroborate earlier work and they suggest that internal waves were the mechanism of cross-shelf transport for *C. magister* megalopae.

Movement of larvae in enclosed fjords system is not well studied, but most work demonstrates that recruitment is highly variable. In Alaska, using light traps similar to those used by Shanks on the Oregon Coast, Herter and Eckert (2008) found that variations in settlement in the complex fjord systems of Alaska were correlated with tidal and lunar cycles. The large variations in settlement between fjords may be explained by small-scale variations in hydrodynamics. Extending the dataset of Herter and Eckert (2008), Smith and Eckert (2011) found highly variable recruitment at both regional (>300 km) and small scales (2–6 km) in different fjords in Southeast Alaska. They suggest that the variability in recruitment both spatially and temporally can be explained by the complex circulation patterns present in the study area. In the enclosed waters of Puget Sound, Dinnel et al. (1993) tracked juvenile cohorts and reported that enclosed basins appear to rely on self-recruitment, and recruitment from other sources such as the ocean is limited. Additional research is needed to understand the movements and/or retention of larvae in enclosed basins.

In the California Current, most megalopae settle on the continental shelf; however, some migrate into estuaries to settle (Miller and Shanks, 2004). Estuaries have numerous fronts which concentrate larvae and may act as a conduit for the transport larvae into the estuaries from the continental shelf (Eggleston et al., 1998). For tidally generated fronts to transport megalopae into an estuary, megalopae must be concentrated in the tidal prism (the volume of water advected into or out of the estuary by the tide; Roegner et al., 2003). Once megalopae are ready to metamorphose into juveniles, it appears that most megalopae metamorphose under the cover of darkness, and the moulting of one megalopa will induce other megalopae to moult (Fernandez et al., 1994a).

3.3. Adult and juvenile biology

Although most C. *magister* settle on the continental shelf (within 10–15 km of shore), most available information is on the settlement and biology of juvenile crabs inside estuaries (Carrasco et al., 1985; Methot, 1989). Thus, throughout this section, unless stated otherwise, studies on the biology of juvenile crabs occurred in estuarine systems.

3.3.1 Habitat

Many studies suggest that shell habitat (more specifically oyster beds) is important for the survival of juvenile crabs. However, shell habitats similar to oyster beds are not common in most Pacific Northwest estuaries or on the continental shelf and the majority of settlement likely occurs in open habitats

(Dumbauld, 1993). The large number of studies examining the role of shell habitat is due to attempts by the Army Corp of Engineers to mitigate the effects of dredging shipping channels in estuaries (Iribarne et al., 1995). In areas without significant shell deposits, gravel/rocky habitats covered with macroalgae and eelgrass (*Zostera marina*) beds have the highest concentrations of juveniles (McMillan et al., 1995).

Adult Dungeness crabs live in coastal regions including the continental shelf, small estuaries and extensive inland waters (e.g. Puget Sound and Southeast Alaska) at water depths ranging from the intertidal to approximately 230 m (Jensen, 1995). In Puget Sound, based on observations from a submersible, most non-ovigerous females were distributed at depths from 20 to 80 m, while males were distributed at depths from 10 to 20 m (Armstrong et al., 1988). In an Alaskan fjord, the use of acoustic tags demonstrated that both males and females reside at depths >40 m during the winter and moved into shallow nearshore waters, <8 m, during the spring when larvae are released (Stone and O'Clair, 2001). Adults are primarily found in sandy-mud bottoms (Cleaver, 1949), where they bury into the sediment and possibly bury on a circadian rhythm, most commonly emerging from the substrate during nocturnal high tides (McGaw, 2005; Stevens et al., 1982).

3.3.2 Movement

Tracking juvenile cohorts in an estuary demonstrated that many 1-year-old crabs migrated out of the estuary onto the continental shelf, and by 2 years of age, all juvenile crabs had migrated onto the continental shelf (Collier, 1983; Stevens and Armstrong, 1984). Tagging studies of adult males and females in the open ocean off Northern California and Oregon have found that distances travelled over nine months ranged from ~0.2 km to as great as ~100 km, but the majority of adults move less than 20 km (Cleaver, 1949; Diamond and Hankin, 1985; Hildenbrand et al., 2011; Snow and Wagner, 1965; Stone and O'Clair, 2001, 2002; Waldron, 1958). The average daily movement was 1.1–3.2 km day^{-1}. Prior to spawning, many crabs move into the nearshore and/or estuaries (Barry, 1985). This does not imply that mating and larval release only occurs within estuaries, rather most larval release likely occurs on the continental shelf. Most female movement is across the continental shelf, but for males, most movement occurs in the alongshore direction (Diamond and Hankin, 1985; Hildenbrand et al., 2011). In the inland waters of British Columbia, research suggests that males retreat to greater depth during winter than females, but overall, female crabs are more active over the course of a year (Smith and Jamieson, 1991a).

In estuaries, many populations move in and out of the intertidal each day to forage. Intertidal foraging is necessary to account for the extreme energy requirements of the large number of individuals present in estuaries (Holsman et al., 2003). Most migration into the intertidal environment occurs under the cover of darkness, so crabs can avoid visual predators (Holsman et al., 2006). If a preferred prey source of adults is present in the intertidal (and crabs are close enough to migrate), they will migrate into the intertidal to forage, even though their need to osmoregulate dramatically decreases the rate of digestion (Curtis et al., 2010; Stevens et al., 1982). Individuals must osmoregulate since salinities and oxygen level in the intertidal are lower from that of subtidal waters were *C. magister* usually inhabit. Thus, many individuals that forage in the intertidal must retreat to depth in order to digest their food.

3.3.3 Diet

Dungeness crabs are opportunistic feeders that are highly adapted to feeding in sandy habitats and do not appear to display strong preferences for specific prey items (Lawton and Elner, 1985). Juvenile *C. magister* are omnivorous and estuarine populations are able to capture more prey than continental shelf populations, which likely explains the increased growth rate of juveniles in estuaries (Jensen and Asplen, 1998; Tasto, 1983). Near San Francisco Bay, stomach contents of crab caught in an estuary had more bivalves in them and juveniles caught in the ocean had more fish (Tasto, 1983). However, in Washington, 1-year-old individuals mostly had crustaceans and mollusks in their stomachs, and 2-year-olds had high concentrations of crustaceans and fish (Stevens et al., 1982). The variety of food sources in these studies corroborate the opportunistic feeding of *C. magister*. This strong ontogenetic shift in feeding patterns of *C. magister*, as shown in Washington estuaries, may minimize competition and cannibalism between cohorts. However, there is no evidence of seasonal cycle in feeding (Stevens and Armstrong, 1984; Stevens et al., 1982).

Diets of adult *C. magister* have been closely examined, and overall, bivalves appear to be the most important food source (Butler, 1954; Gotshall, 1977; Stevens et al., 1982). However, the three studies just referenced found that different food items were most prevalent in the stomach contents: Butler (1954) clams, Gotshall (1977) fish and Stevens et al. (1982) Crangon shrimp. All of these studies, however, also found high concentrations of bivalves in the stomachs of crabs. *C. magister*, especially females, are well known to be highly cannibalistic on recently moulted juveniles (Botsford and Hobbs, 1995; Eggleston and Armstrong, 1995; Fernandez, 1999; Stevens et al., 1982).

Adult crabs feed by probing the substrate with their claws (chelae) until a prey item is detected, at which point they contract their claws and remove the food. Adult *C. magister* have been observed excavating heart cockles (*Clinocardium nuttallii*; Butler, 1954). In the laboratory, bivalves buried in artificially oiled sediments were unable to bury as deep and thus were consumed more frequently by *C. magister* (Pearson, 1981). *C. magister* are also able to detect (at a distance) ground-up clams frozen in sea water using chemosensory abilities at concentrations of $10^{-10}\,g\,l^{-1}$ of clam extract (Pearson, 1979). In laboratory studies, crabs preferentially consumed smaller clams when given a choice (Juanes and Hartwick, 1990). Consumption of larger clams increased the probability of claws being damaged when cracking open clams, and crabs with damaged claws were unable to crack open clams.

3.3.4 Cannibalism

Cannibalism by young-of-the-year crabs on newly settled megalopae can be extremely high (Armstrong et al., 1988; Dumbauld, 1993; Eggleston and Armstrong, 1995). Juvenile crabs are highly cannibalistic and researchers hypothesize that moulting (from megalopae to juvenile and between juvenile instars) under the cover of darkness minimizes cannibalism. Cannibalism by the first settlement cohort strongly influences survival of later cohorts (Fernandez et al., 1993). In one study, as population density of young-of-the-year crabs increased, the total number of juveniles consumed by cannibalism increased, but the likelihood of juveniles being eaten decreased proportionally (Fernandez, 1999). Additionally, as the density of crabs increased, individuals emigrated away from areas of high density even when food abundance was artificially enhanced (Iribarne et al., 1994). This suggests that density-dependent cannibalism may strongly influence population dynamics.

3.3.5 Predation

Juvenile *C. magister* are common food items for a multitude of predators such as starry flounder (*Platichthys stellatus*), English sole (*Parophrys vetulus*) and the Staghorn sculpin (*Leptocottus armatus*), probably the most significant predator in estuaries (Armstrong et al., 1995, 2003). Juvenile *C. magister* are also consumed by adults of the introduced European Green crab (*Carcinus maenas*), although their habitats (vertical range in the intertidal) do not overlap, and thus, predation pressure is minimal (McDonald et al., 2001). Due to the large size of adult *C. magister*, they have relatively few predators. Well-known predators of adult crabs are lingcod (*Ophiodon elongatus*), Cabezon (*Scorpaenichthys marmoratus*) and wolf eel (*Anarrhichthys ocellatus*; Reilly, 1983c). The primary habitat of these three fish species (rocky bottom) and *C. magister* (sandy bottom) often does not overlap. In Southeast Alaska, crabs make up ~15% of the diet of sea otters (*Enhydra lutris*) (Garshelis et al., 1986). Repeated test fisheries after the introduction of otters reported a 61% decline in the abundance of adult *C. magister* in areas where otters were prevalent. In recent years, where sea otter populations have recovered, the test fishery catch was not significantly different from zero for pots fished in <60 m depth of water (Shirley et al., 1996). Depths >60 m are likely a predator refuge for *C. magister* since most otters do not dive to depths >60 m to forage (Bodkin et al., 2004). In a nearby estuary where otters were not present, test fishery catch was significantly higher than catch where otters were present. These data suggest that otters strongly influence the location and depth at which adult *C. magister* reside.

3.3.6 Competition

Although many studies report increased juvenile abundance in habitats created with shell hash, these studies only assessed juvenile abundance in the years directly following the creation of the habitat. Recent work has shown that after multiple years, the shell habitat is colonized by adult Hairy (Yellow) Shore Crabs (*Hemigrapsus oregonensis*) which have a strong negative effect on *C. magister* recruitment, reducing recruitment of *C. magister* to almost zero (Visser et al., 2004). *H. oregonensis* are capable of outcompeting juvenile *C. magister* for food and evicting them from refuges; however, they inhabit the high–low intertidal, so there is relatively little overlap of the two species habitats. *C. maenas* is also a stronger competitor than juvenile *C. magister* and outcompetes *C. magister* in nocturnal feeding trials. Additionally, *C. maenas* causes *C. magister* to emigrate away from 'higher-quality' habitat (McDonald et al., 2001). However, in Washington, where the studies were conducted, the habitat of *C. magister* and *C. maenas* does not currently overlap, so competition is minimal. In addition to interactions with other crab species, conspecific interactions have demonstrated that first and second juvenile instars are less aggressive towards other stages than later stages, with stages 3–6 being the most aggressive (Jacoby, 1983). Additionally, interactions between adult males and females often result in females submitting to males (Jacoby, 1983).

3.3.7 Growth and development

The carapace width of newly settled juveniles is approximately 5 to 7 mm (Butler, 1961). *C. magister*, like other crustaceans, grow by moulting their shell (Jaffe et al., 1987; Ruppert, 1994). Unlike adults, juvenile *C. magister* moult multiple times over the course of their first (~6 times) and second years (~4 times), which allows them to grow rapidly (Tasto, 1983). Within estuaries, the growth rate of juveniles is much faster than for juveniles that settle on the continental shelf; by the end of their first summer, estuarine crabs are ~40-mm carapace width, while oceanic juveniles are ~10-mm carapace width (Gunderson et al., 1990). However, by comparing reported carapace width to those observed in ROV videos during extraordinarily high recruitment years, Shanks et al. (2010) demonstrated that the carapace width of crabs was significantly smaller in years of high recruitment.

For adults, moulting occurs annually during a relatively short time period of 6–8 weeks (Hankin et al., 1989; Mohr and Hankin, 1989). Prior to moulting, crab shells are often heavily fouled with barnacles and other sessile organisms, whereas crabs are free of fouling after moulting (Cumbrow, 1978). Just prior to moulting, a suture line forms where the shell will open and the crab will back out of the old exoskeleton. Moulting occurs from March to June in California, April to September in the San Juan Islands, and June to July in Alaska (Table 3.1; Jaffe et al., 1987; Knudsen, 1964; Park and Shirley, 2008; Wild, 1983b). Approximately, 90% of female crabs with a carapace width of ~135 mm moult annually, while almost no female crabs with a carapace width of >155 mm moult annually (Hankin et al., 1989).

Adult crabs gain between 8.1- and 19.7-mm carapace width following each moult (Hankin et al., 1989). After moulting, it takes approximately 2 months for the exoskeleton to completely refill with tissue. Miller and Hankin (2004) provide descriptions of individual moult stages for determining moult stage of crabs in the laboratory; Washington Department of Fish and Wildlife also conducts routine field surveys in Puget Sound and collects extensive moult status using *in situ* tests based on shell hardness. Department of Fish and Oceans in British Columbia uses a durometer to measure shell hardness and defines soft shells as shells under 70 units (Canada. Dept. of Fish. and Oceans. Pacific Region., 2012).

3.4. Mortality (all stages)

3.4.1 Natural mortality

Estimates of intra-annual mortalities of larvae are predicted to be consistent and the average daily survival rate is 0.066 day^{-1} (Hobbs et al., 1992). This calculation is based on daily survival of larvae from plankton tows collected in the upper portion of the water column. As larvae migrate to depth each day, it is likely that their sampling missed a large percentage of larvae and thus underestimated daily survival. Annual mortality rates of adults have been estimated at 2.5% for sublegal males and 1.3% for females, though the results are controversial due to the statistical techniques used (Butler and Hankin, 1992; Smith and Jamieson, 1989a, 1991a, 1992). Models based on metabolic rates of adult crabs suggest that their lifespan ranges from 8 to 10 years (Gutermuth, 1989).

3.4.2 Diseases

In the laboratory, zoea are highly susceptible to infection (e.g. *Lagendium* sp.) and need to be reared with fungicides and antibiotics to prevent infections (Armstrong, 1976; Caldwell et al., 1978; Fisher and Nelson, 1977, 1978). Adult *C. magister* can have a multitude of diseases and parasites such as microsporidia in skeletal muscles, systemic ciliates and trematodes in the nervous and connective tissue (Morado and Sparks, 1988). Although most of these diseases have not been reported to have detrimental effects on the adult population, one *Chlamydia*-like bacteria may have caused mass mortalities in crab pots and holding tanks in Willapa Bay (Sparks et al., 1985). The microsporidia *Nadelspora canceri* infects crabs in the California Current (Childers et al., 1996), and populations in small estuaries along the coast were more heavily infected than offshore populations and populations in Puget Sound and Glacier Bay, Alaska, were not infected. Although infections do not appear to influence patterns of abundance in *C. magister*, they could potentially have an effect on populations in confined systems such as aquaculture or flow-through tanks for resale.

3.4.3 Pesticides

Pesticides can kill *C. magister*. For instance, the insecticide Sevin®, which is commonly used in oyster culture, prevented eggs from hatching at a concentration of 1 mg l^{-1} and killed 50% of zoea at a concentration of 0.01 mg l^{-1} (Buchanan, 1970). When adult crabs consumed cockles that had been exposed to the insecticide Sevin at 10 mg l^{-1}, all were irreversibly paralyzed and 77% of crabs were paralyzed when they consumed clams that had been maintained at a concentration of 3.2 mg l^{-1} of the insecticide Sevin (Buchanan, 1970). Fifty percent of adults exposed to Sevin at a concentration of 0.26 mg l^{-1} died within 24 h of exposure. The insecticide methoxychlor was tested at much lower concentrations than Sevin, and 50% of adults died after exposure for 96 h at a concentration of 130 µg l^{-1}. In those that survived, the chemical was concentrated in their exoskeleton (Armstrong et al., 1976). Additionally, multiple different herbicides and heavy metals (e.g. cadmium, copper and mercury) also have lethal effects on the larvae of *C. magister* (Caldwell et al., 1979; Martin et al., 1981).

3.4.4 Dredging

Multiple studies have examined the effects of dredging the Gray's Harbor, WA shipping channel on *C. magister* and assessed potential ways to mitigate the effects of dredging (e.g. creating oyster shell habitat) (Armstrong et al., 1987). The effects of hopper, pipeline and clamshell dredges have all been examined; however, hopper dredges are the most commonly used, and therefore, I only report results from studies on this form of dredging. The average hopper dredge entrains between 0.046 and 0.587 juvenile crabs with each cubic yard of sediment. Of the crabs that were entrained, 86% with a carapace width of >50 mm died and 46% of crabs with a carapace width of <50 mm died. Adults captured in dredges that were not killed were able to dig out of sediments less than 20 cm deep (Chang and Levings, 1978).

3.4.5 Hypoxia

Another threat to *C. magister* population is hypoxia. Recent die-offs of adult *C. magister* observed off the coast of Oregon have been attributed to low dissolved oxygen (hypoxia) events (Chan et al., 2008). Laboratory studies have examined feeding rates and behaviours of adult *C. magister* in hypoxic conditions and determined that crabs cease feeding below 3.2 kPa O_2 (Bernatis et al., 2007; McGaw, 2008). These researchers suggest that reduced feeding lowers the number of physiological processes that occur and minimizes oxygen consumption. Thus, consuming more food prior to entering the hypoxic regions likely increases the survival of crabs. In an estuary in British Columbia, using acoustic tags equipped with CTDs, researchers found that crabs actively avoided areas of lower salinity and have behaviours (not described) that aid them in avoiding and surviving hypoxic conditions (Bernatis et al., 2007; Curtis and McGaw, 2008). In a seasonally hypoxic fjord, researchers used acoustic tags to determine whether crabs would migrate into the shallow nearshore environment or north to avoid the hypoxic region of the fjord (Froehlich et al., in review). Their results demonstrated that crabs migrated into the shallow nearshore environment rather than northwards out of hypoxic waters.

3.4.6 Egg predation

Within the egg masses of *C. magister*, there are often predatory nemertean worms, *Carcinonemertes errans*, which consume developing eggs (Wickham,

1978, 1979b). The worms can consume approximately 5 eggs worm^{-1} day^{-1} (Wickham, 1980). Throughout the range of *C. magsiter*, both juvenile and adult worms cover the surface of adult crabs (Wickham, 1979a). The majority of work on *C. errans* is based on adult crabs that were collected in the open ocean even though *C. magister* is well known to inhabit estuaries throughout its range. Recent work has demonstrated that the level of infection decreases as crabs move further up estuaries into lower salinity waters, suggesting that estuaries provide adults *C. magister* with a refuge from *C. errans* (Dunn, 2011). One would expect that the decrease in infestation would be due to decreases in salinity; however, mortality studies showed that *C. errans* were able to tolerate similar temperatures and salinities as adult *C. magister*, which suggests that some other factor causes infestation to decrease along an estuarine gradient.

3.4.7 Climate change

As the climate continues to change, researchers have hypothesized that increasing ocean temperatures will cause a northwards movement of predators and competitors of *C. magister* (McConnaughey and Armstrong, 1995). Additionally, they suggest that *C. magister* will likely inhabit greater water depths and release their larvae earlier in the year. Early work on ocean acidification suggests that there will be minimal impacts on adult *C. magister* since the adults are able to recover their haemolymph pH after exposure to acidic waters (Pane and Barry, 2007; Ruttimann, 2006). Work on the effects of acidification on the development of *C. magister* larvae suggests that there will be few effects on development (R. Descoteaux, personal communication). Research seems to demonstrate that the wide range of habitats currently occupied by *C. magister* makes the organism relatively plastic, and thus the organism will be able to change habitats or behaviours to cope with climate change.

4. FISHERY
4.1. History

The *C. magister* fishery has been reviewed extensively in other articles and I will only provide a brief overview and emphasize recent management protocols (Demory, 1990; Didier, 2002; Melteff, 1985). The following history of the fishery is adapted from Dahlstrom and Wild (1983) and augmented to include more recent findings.

Tribes along the West Coast of North America consumed *C. magister* (Dahlstrom and Wild, 1983). The Yurok tribe in Northern California is reported to have speared crabs, while other tribes gathered them by hand, often focusing on young adults and juveniles in the nearshore environment (Greengo, 1952; Losey et al., 2004). The non-tribal fishery began in San Francisco Bay in the 1860s using hoop nets equipped with cedar, cork or copper floats. The first reported annual commercial catch occurred in the late 1840s, and in 1863, the California Department of Fish and Game recorded the first landing. In 1897, due to observed declines in the San Francisco Bay population, a moratorium was placed on retaining female crabs. The hope was that releasing female crabs would minimize the effect of the fishery on reproductive output. Subsequently, San Francisco fishermen requested seasonal closures and the first seasonal closure occurred in 1903. The fishery was closed from September 2–October 31 to avoid catching soft-shelled crabs. In California, the first size restriction was initiated in 1905; that is, crabs had to be a minimum size of 6″ or 152 mm. Small sailing vessels that were capable of operating approximately 50 hoop nets a day dominated the early fleet. In the early 1900s, gasoline engines started to be more common, and vessels were able to operate as many as 100 hoop nets each day. The number of participants in the fishery dramatically increased in the 1930s when crab pots were introduced.

In Oregon, the first Dungeness crab landings occurred in 1889 (Demory, 1990). The first seasonal closures in Oregon occurred during the 1948–1949 season in order to minimize the retention of crabs in poor condition (low yield of meat), but the timing of the closure was different north and south of Cascade Head to account for latitudinal variation in the timing of moulting. Additionally, at this time, the first closure to the retention of female crabs occurred. From 1909 to 1933, commercial fishermen had daily and/or annual catch limits. When catch limits were repealed, annual landings increased dramatically (Waldron, 1958). In 1996, Oregon established a limited-entry programme to prevent a large number of boats from entering the fishery when catch levels are high. I was unable to find historical reviews of the fisheries in Washington, British Columbia and Alaska. It should be noted that in Washington, the 1994 federal court order known as the Rafeedie Decision stated that Washington Treaty tribes had the right to shellfish under their treaties and thus Dungeness crabs have been co-managed by the tribes and state since.

Landings in the California Current fishery (California, Oregon and Washington) continued to rise until 1948 at which point catch began to oscillate (Figure 3.3; Demory, 1990). Early on, these oscillations were decadal in cycle, but since 1980s, the decadal pattern to the oscillation is no longer present (Figure 3.3; Shanks and Roegner, 2007). It is well known that in the late 1970s, fleet size (and effort) increased dramatically with the creation of exclusive economic zones, which coincides with the end of the oscillations (Figure 3.4; Gelchu and Pauly, 2007; Shanks and Roegner, 2007). Further it has been estimated that historically as little as 40% of the legal males were extracted annually and following the fleet expansion in 1970s >90% of legal males have been/are extracted annually (Methot and Botsford, 1982). Based on models developed by Botsford et al. (1983), Shanks and Roegner (2007) suggests that during periods when effort was low some crabs escaped and contributed to the next year's fishery, but in years with increased effort, effectively all legal crabs were extracted. Thus, he hypothesizes that the early oscillations in catch may have been due to increases and decreases in effort based on the relative abundance of legal crabs.

In the Alaska Current, catch in British Columbia and Alaska has fluctuated but not on a decadal cycle like that of the California Current (Figure 3.3). In the Alaska Current, the fishery primarily occurs in inland waters where larval dispersal is influenced by complex hydrodynamics that likely lead to area of larval retention. I hypothesize that the effect of the complex hydrodynamics on larval dispersal has contributed to the fluctuations and crashes in certain fisheries; however, limited research has been conducted and future work should examine the dispersal of larvae in these complex hydrodynamic environments (Orensanz et al., 1998).

Despite the fluctuations, in 2011, *C. magister* accounted for only ∼5%, 6% and 15% of the total biomass harvested in California, Oregon and Washington but accounted for 25%, 30% and 44% of the total revenue (Figures 3.5–3.7). These numbers demonstrate that although Dungeness crab does not account for the greatest biomass harvested it is the most economically important species harvested (second to squid in California) in the California Current. As the fishery has become more lucrative, the fishery is essentially a race fishery with most landings occurring within the first 2 months of the fishery opening, which floods the market with crab (Hackett et al., 2003). In other fisheries, extending catch over a longer time period has increased the profit margin of the fishery; however, an economic analysis of the California fishery surprisingly found that there would likely be little profit increase by extending the fishery (Hackett et al., 2003). The researchers determined that frozen-picked crabmeat is the most profitable and, since it is frozen when it is caught, does not increase the value.

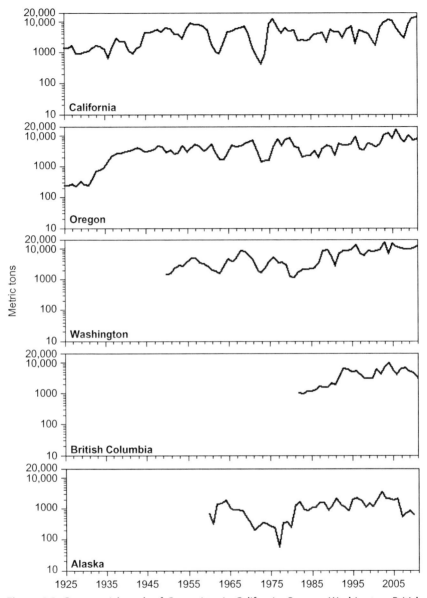

Figure 3.3 Commercial catch of *C. magister* in California, Oregon, Washington, British Columbia and Washington. Catch data were not available back to 1925 in all regions. Note the decadal patterns (significant autocorrelation) in catch from 1950s to 1980s in commercial catch in California, Oregon and Washington (Shanks and Roegner, 2007). This decadal has not occurred since 1980s (no further autocorrelation).

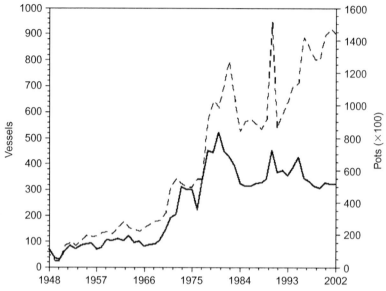

Figure 3.4 Number of boats and pots participating each year in the commercial *C. magister* fishery in Oregon. Note the dramatic increase in the effort (both pots and vessels) that occurred in the mid 1970s. This timing coincides with the federal plan to enhance U.S. fisheries following the establishment of the Exclusive Economic Zone (Gelchu and Pauly, 2007). Data on effort were spotty and thus are not presented here.

However, the race fishery increases the number of pots on the fishing ground and causes fishermen to take extra risks often leading to loss of life (Dewees et al., 2004). A survey of fishermen in California found that the only 2 (out of 12) accepted options for stretching out the season are either fixing a trap limit regardless of vessel size or restricting fishing to daylight hours, though the authors note that in other regions where similar restrictions have been implemented, the number of pots has not decreased (Dewees et al., 2004).

4.2. Management

The fishery for *C. magister* is managed using the 3-S management technique. The 3-S management technique controls the sex of individuals that are harvested, the minimum size of the individuals that are harvested and the season when harvesting occurs (Table 3.4). Didier (2002) provides excellent comparative tables of the regulations for California, Oregon and Washington and I include only a small portion of these tables here (Table 3.4). Commercial fisheries throughout the range of *C. magister* are only allowed to

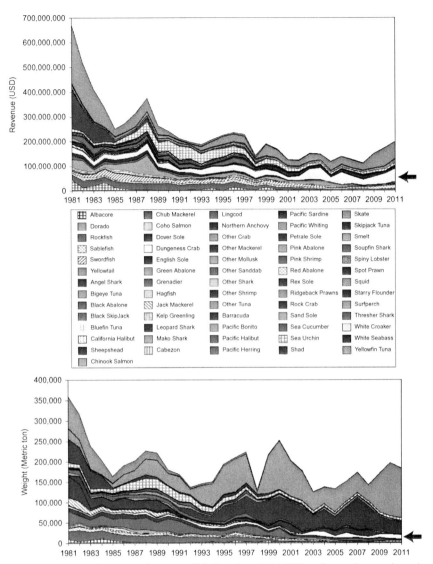

Figure 3.5 Annual revenue (top panel) inflated to 2012 USD value and annual catch (bottom panel) for commercial fisheries in California from 1981 to 2011. (Pacific Fisheries Information Network (PacFIN) retrieval dated December 2012, Pacific States Marine Fisheries Commission, Portland, Oregon (www.psmfc.org).) Arrows on the right side of the figure point to the area that represents *C. magister*. (For colour version of this figure, the reader is referred to the online version of this chapter.)

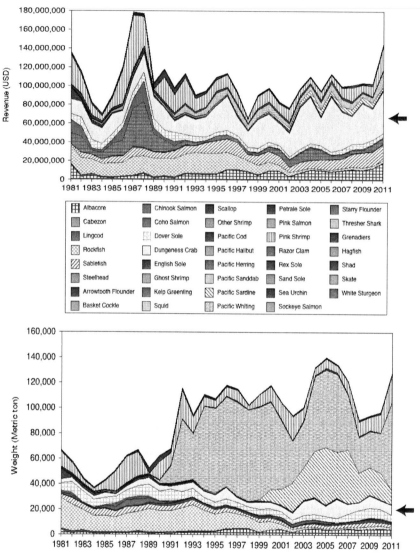

Figure 3.6 Annual revenue (top panel) inflated to 2012 USD value and annual catch (bottom panel) for commercial fisheries in Oregon from 1981 to 2011. (Pacific Fisheries Information Network (PacFIN) retrieval dated December 2012, Pacific States Marine Fisheries Commission, Portland, Oregon (www.psmfc.org).) Arrows on the right side of the figure point to the area that represents *C. magister*. (For colour version of this figure, the reader is referred to the online version of this chapter.)

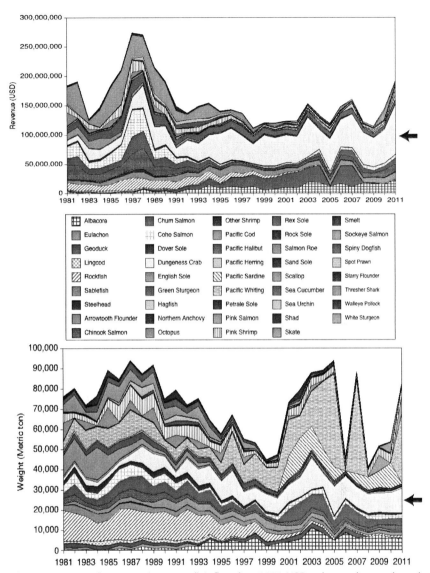

Figure 3.7 Annual revenue (top panel) inflated to 2012 USD value and annual catch (bottom panel) for commercial fisheries in Washington from 1981 to 2011. (Pacific Fisheries Information Network (PacFIN) retrieval dated December 2012, Pacific States Marine Fisheries Commission, Portland, Oregon (www.psmfc.org). Arrows on the right side of the figure point to the area that represents *C. magister*. (For colour version of this figure, the reader is referred to the online version of this chapter.)

Table 3.4 Sport and commercial fishery retention size, season and unique regulations to each region throughout the range of *Cancer magister*

	Legal size (mm)	Season	Notes
Sport fishery			
California	145	November 5–July 30	Limit 10, females can be retained
Oregon	145	Bays Open Year Round, Ocean December 1–September 30	Limit 12
Washington: Coast	152	December 1–September 15 for pots, year round all other gear	Limit 5 southern coast, Limit 6 northern coast
Washington: Puget Sound	160	July 1–September 3 (Thursday–Monday)	Limit 5 (many small regional closures)
British Columbia	165	Year round	Limit 4–6 (many small regional closures)
Alaska	165	Year round	Limit 20 crabs (multiple regional regulations)
Commercial fishery			
California	160	December 1–June 30 (varies North to South)	
Oregon	160	Bay fishery open weekdays January–Labour Day Ocean Fishery Open December 1–August 15	Pot limits allocated by historic catch
Washington: Coast	160	December 1–September 15	Pot limits allocated by historic catch
Washington: Puget Sound	160	October 1–April 15	100 pots per permit
British Columbia	165	Due to the complex hydrodynamic regulations are variable	Number of pots variable based on region
Alaska	165	Due to the complex hydrodynamic regulations are variable	

harvest male crabs, though this is a recent change in British Columbia where it used to be legal to retain females (Table 3.4). For recreational fisheries, all states except California restrict catch to only male crabs. The size of individuals that can be retained in commercial and sport fisheries varies throughout the crabs range from 145 to 165 mm (Table 3.4). Most research suggests that crabs retained in the fishery are ~4 years of age (Botsford, 1984). However, models of growth rates based on water temperatures suggest that the range could be from 2 to 8 years for crabs to enter the fishery with longer time periods occurring where average water temperatures are lower (Gutermuth, 1989). The commercial season is set to eliminate fishing while crabs are moulting and, due to the variability in the timing of moulting, the dates of closure change with latitude with openings occurring later further north (Tables 3.1 and 3.4). In the California Current, the season is delayed if the percentage of meat/body weight is not above 25–30%. At the northern extent of the fishery, in Puget Sound, inland waters of British Columbia and Alaska, regulations are complex due to the convoluted coastline and stocks are managed as multiple subunits. While in the California Current the population is well connected, it is likely that the complex hydrodynamics of inland waters reduces connectivity between populations ultimately making them into separate stocks. The recreational season varies throughout the range of *C. magister* and also changes with gear type (e.g. hoop net vs. pot).

4.3. Direct and indirect impacts of the fishery
4.3.1 Ghost fishing
Each year a large number of crab traps are lost; in Puget Sound, an estimated 12,193 pots are lost in the commercial and recreational fisheries combined, and in British Columbia, an estimated 11% of traps are lost each year (Antonelis et al., 2011). Due to their effective design, crab pots have a propensity to continue to capture crabs (ghost fish) after they are lost. In Puget Sound, 72 h after pots were returned to the water with their original catch, 79% of legal crabs, and 33% of sublegals remained in traps (High, 1976). Over the course of a year, it was estimated that as much as 7% of the annual catch is harvested by ghost fishing (Breen, 1985a,b). An examination of the economic value of crabs caught and killed by ghost traps ($37–$91 pot^{-1}) and compared to the cost of removing derelict gear ($93–$193 per pot) suggests that there is usually little economic incentive for removing derelict pots (Antonelis et al., 2011). It is becoming a requirement throughout the California Current for the lids of pots to be secured with cotton twine that will rot away when pots are lost. Thus, cord attached to the enclosure

that rots away in seawater should be reduced in diameter so that it will rot within 50 days and subsequently open the door to the pot. Additionally, having pots with doors that open on the top of the pot may not allow trapped crabs to escape even after the clasp breaks since the lids are often held closed by encrusting organisms. Crabs in pots (and more specifically small crabs) have an increased probability of being injured as soak time (the time the pot is in the water) increases (Barber and Cobb, 2007). However, preventing crabs from utilizing their pinchers did not decrease in pot mortality, suggesting that cannibalism within pots is not common (Shirley and Shirley, 1988).

4.3.2 Handling mortality
Sixteen percent of soft-shell crabs die after being handled only once, while only 4% of hard-shell crabs die from a single handling event (Tegelberg, 1970). By tagging individuals and returning them to pots in the water, Tegelberg (1971) determined that mortality of soft-shell crabs was 10% after 2 days and 25% after 7 days. If crabs were handled three times in 6 days, handling mortality increased to 41%. These findings of mortality on soft-shell crabs were corroborated in additional studies off the Washington coast (Barry, 1981).

4.3.3 Trawl fishery
The indiscriminate nature of many benthic trawl fisheries has led researchers to speculate that by-catch of *C. magister* is high in benthic trawl fisheries. A study in California near the Farallon Islands reported a mortality of 0.53 male crabs per hour of trawling and a mortality of 0.12 legal-sized male crabs per hour of trawling (Reilly, 1983b). All sluggish crabs caught in the trawl that were held in flowing seawater for 3–20 h fully recovered. However, these estimates are likely low as the study was conducted off San Francisco (a region with a relatively small population). This chapter references (without citation) a study off Washington that reported trawling induced mortalities of ~4.2% for both sexes. NOAA observer data from 2008 reported that ~387 metric tons of *C. magister* were discarded in both the limited entry bottom trawl fishery and California halibut (*Paralichthys californicus*) bottom trawl fisheries in California, Oregon and Washington (Bellman et al., 2010). In 2008, a total of 19,899 metric tons were landed by the crab fishery in California, Oregon and Washington and thus by catch from the trawl fisheries only accounted for ~2% of the total catch. A comparison between a mark-recapture study where crabs were caught with an otter trawl and a study where crabs were caught with crab pots

determined that recovery rates were equal between the two studies and it was argued that this demonstrates that trawl fisheries have a minimal effect on *C. magister* (Anonymous, 1949). However, no data to substantiate this claim are provided in the report.

4.3.4 Impact to benthos

To my knowledge, no work has examined the effect of *C. magister* traps on the benthos. In British Columbia, a study examined the impact of Spot Prawn (*Pandalus platyceros*) pots on benthic communities with a primary focus on damage to sea whips (*Halipteris willemoesi*) (Troffe et al., 2005). They found that in 600 hauls, 30 sea whips were brought to the surface and of these 50% were damaged. A project over 4 years compared areas fished with pots to soft-sediment areas not fished with pots and demonstrated that there was no difference in the benthic communities (Coleman et al., 2013). A study of traps deployed on reefs demonstrated that there were significant impacts on the benthos, especially when wind caused the traps to be moved across the reef (Lewis et al., 2009). Research seems to suggest that pots have a minimal effect on soft sediment communities; however, the ecological impact of traps on benthic communities is a topic that needs extensive research in the future.

4.4. Fishery prediction

Due to the economic importance of *C. magister* and historic fluctuations in catch, many research projects have attempted to predict commercial catch the year prior to a fishery and predict the cause of large-scale population variations observed throughout the range of the species.

4.4.1 Catch prediction

By using crab pots modified to retain juvenile crabs, researchers are able to predict commercial catch one year in advance with 10–20% accuracy; however, to my knowledge, this has not been implemented by any state (Stefferud, 1975). In Puget Sound, standard crab pots with the escape rings closed are used to help predict fisheries (Fisher and Velasquez, 2008). Since soak times vary between pots, Smith and Jamieson (1989b) presented a statistical model that can be used to standardize catch between traps with different soak times. In a comparison of pots and SCUBA methods for predicting catch, pot surveys were more effective except in time periods when crabs were moulting and, hence, not actively moving (Taggart et al., 2004). Gunderson and Ellis (1986) developed a modified plumb staff beam trawl that is highly effective at sampling juvenile *C. magister*.

McConnaughey and Conquest (1993) examined data collected during these trawl surveys and concluded that geometric means are a better estimator of abundance than arithmetic means. A comparison of towed camera sleds and trawls demonstrated that trawls routinely underestimated the abundance of adult crabs (Spencer et al., 2005). Although there appear to be methods that are successful for predicting fisheries, to my knowledge, none of these procedures have been implemented and thus their validity cannot be assessed.

Shanks and colleagues (2010, 2007) have been using light traps to capture the megalopae of *C. magister* (see earlier discussion in Section 3.2). Using the amount of megalopae caught annually, they have been able to predict commercial catch 4 years later (the time it takes megalopae to grow into commercial-sized crabs) with an accuracy of $\sim 12\%$. In recent years, catch of megalopae has increased to the point that their early linear model would have predicted a commercial catch of $\sim 700,000,000$ lbs (10 times greater than historic maximum). Commercial catch correlated with these high megalopae returns has levelled off with (and declined; Figure 3.8; Shanks, in press). In years when a large number of crabs recruit to the population, density-dependent effects increase, causing commercial catch to level off and decline. Below 100,000 returning megalopae, the population is recruitment limited, and above 175,000 returning megalopae, density-dependent effects cause the population to decrease. This predicts that the greatest commercial catch should occur when $\sim 175,000$ megalopae recruit to the light trap. Although the predictive curve has changed to a second-order polynomial (for everywhere except Central California), the predictive power of the curve is still highly significant (Figure 3.8).

4.4.2 Fluctuation predictions (California Current)

Early on commercial catch of *C. magister* historically oscillated on nearly a decadal cycle; however, the cycles have recently become less regular or have disappeared altogether (Figure 3.3). As discussed earlier, it is likely that these oscillations ceased due to a dramatic increase in effort following the creation of exclusive economic zones. However, commercial catch does fluctuate (not decadally) and considerable research has attempted to explain these fluctuations.

Although the species is fished extensively, the fluctuations are not induced by the fishery mortality or impacted mating success (Botsford et al., 1983; Hankin et al., 1997; McKelvey et al., 1980). McKelvey et al. (1980) generated numerous multistage recruitment models and argued that factors influencing the early egg and/or larval stages cause the variation in

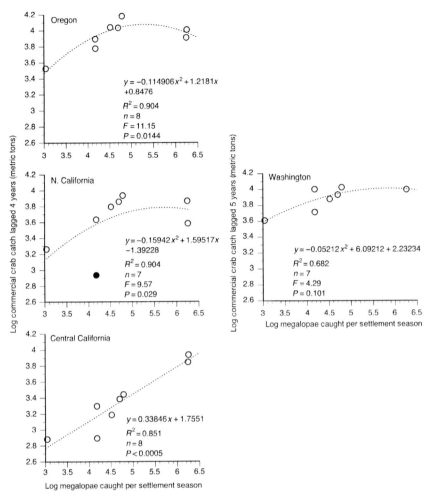

Figure 3.8 Log catch of *C. magister* megalopae caught annually versus commercial catch in Central California (south of Sonoma Country), Northern California (Sonoma County north), Oregon and Washington (Shanks, in press). Commercial catch is lagged 4 years after settlement season except for Washington where it is lagged by 5 years. Dotted lines and statistics are the results of regressions. The filled circle in Northern California is an outlier that was excluded from analysis.

commercial catch. Using a data set collected over 12 years, Shanks (in press) demonstrated that relationship between larval success and adult population size varies with the amount of larval success. It is relatively widely accepted by researchers who suggest that larval success is influenced by hydrodynamics (summarized in Table 3.3 and reviewed earlier when discussing dispersal patterns).

Density-dependent factors alone have not been able to explain the variations in commercial catch (summarized in Table 3.3). Predation on larvae by salmonids and consumption of eggs by *C. errans* do not cause the observed fluctuations (Botsford et al., 1982; Hobbs et al., 1992). Some researchers have suggested that cannibalism on recently settled juveniles may cause the cyclic patterns, though the hypothesis is controversial (Botsford, 1981, 1984; Botsford and Wickham, 1978; Botsford et al., 1983; McKelvey and Hankin, 1981). Most research has attempted to explain the fluctuations by examining either physical or biological perturbations, though it is likely that a combination of the two causes the population fluctuations (Higgins et al., 1997). Shanks and colleagues (Shanks, in press; Shanks and Roegner, 2007; Shanks et al., 2010) provide the best explanation for the fluctuations of the population (see earlier for more in depth explanation). Overall, they have demonstrated that the number of larvae recruiting is positively correlated with adult population size, and during years with high larval success, density dependence affects survival of juveniles.

5. CONCLUSIONS

A large body of the literature has been generated over the years on the biology of *C. magister*, and this review provides an extensive review of the biology of *C. magister* and a brief overview of the commercial fishery. Potential impacts of future changes on the environment are only briefly covered, as this literature is still in the process of being published. Although much has been published on *C. magister*, the research has been surprisingly patchy, for example, most juvenile work occurring in the Grays Harbor and Willapa Bay estuaries. There is an apparent difference that is not well understood between the California Current population and the Alaska Current population with the population in the California Current being much more resilient to exploitation. Further, in years with extremely high recruitment, there are no studies on the density-dependent effects that are occurring. Thus, more studies should be conducted throughout the range of *C. magister* in order to allow researchers and managers to understand which characteristics apply to the entire population and which apply to only certain portions of the population.

ACKNOWLEDGEMENTS

I would like to thank the Oregon Dungeness Crab Commission (ODCC) for funding the writing of this chapter as part of their Marine Stewardship Certification process. B. Butler was invaluable in assisting with finding references. Early drafts were greatly improved by comments from A. L. Shanks, C. Pritchard, M. Jarvis, K. Meyer and A. Burgess. N. Rasmuson provided the drawings for the life cycle figure. J. T. Carlton provided advice on the nomenclature of the species.

REFERENCES

Ainsworth, J., 2006. An Evaluation of the Use of Mating Marks as an Indicator of Mating Success in Male Dungeness crabs. M.S., Humboldt State University.

Anonymous, 1949. Drag boat damage on crabs. Fish Commun. Oregon. Res. Briefs 2, 9.

Antonelis, K., Huppert, D., Velasquez, D., June, J., 2011. Dungeness crab mortality due to lost traps and a cost–benefit analysis of trap removal in Washington State waters of the Salish Sea. N. Am. J. Fish. Manage. 31, 880–893.

Armstrong, D., 1976. A mycosis caused by *Lagenidium* sp. in laboratory-reared larvae of the Dungeness crab, *Cancer magister*, and possible chemical treatments. J. Invertebr. Pathol. 28, 329–336.

Armstrong, D., Armstrong, J., Dinnel, P., 1988. Distribution, abundance and habitat association of Dungeness crab, *Cancer magister*, in Guemes Channel, San Juan Islands, Washington. J. Shellfish Res. 7, 147–148.

Armstrong, D., Buchanan, D., Mallon, M., Caldwell, R., Millemann, R., 1976. Toxicity of the insecticide methoxychlor to the Dungeness crab, *Cancer magister*. Mar. Biol. 38, 239–252.

Armstrong, D., Rooper, C., Gunderson, D., 2003. Estuarine production of juvenile Dungeness crab (*Cancer magister*) and contribution to the Oregon-Washington coastal fishery. Estuaries 26, 1174–1188.

Armstrong, D., Wainwright, T., Orensanz, J., Dinnel, P., Dumbauld, B., 1987. Model of dredging impact on Dungeness crab in Grays Harbor, Washington. Fisheries research Institution, University of Washington, School of Fisheries; FRI-UW-8702.

Armstrong, J., Armstrong, D., Mathews, S., 1995. Food habits of estuarine staghorn sculpin, Leptocottus armatus, with focus on consumption of juvenile Dungeness crab, *Cancer magister*. Fish. B-NOAA 93, 456–470.

Austin, J.A., Barth, J.A., 2002. Drifter behavior on the Oregon–Washington shelf during downwelling-favorable winds. J. Phys. Oceanogr. 32, 3132–3144.

Bakun, A., 1996. Patterns in the Ocean: Ocean Processes and Marine Population Dynamics. California Sea Grant Program, NOAA, La Paz, Mexico.

Barber, J.S., Cobb, J.S., 2007. Injury in trapped Dungeness crabs (*Cancer magister*). ICES J. Mar. Sci. 64, 464–472.

Barry, S., 1981. Coastal Dungeness crab study. State of Washington Department of Fisheries, Project Number 1-135-R (3).

Barry, S., 1985. Overview of the Washington coastal Dungeness crab fishery. In: Melteff, B. (Ed.), Proceedings of the Symposium on Dungeness Crab Biology and Management. University of Alaska Sea Grant, Fairbanks.

Bellman, M., Heery, E., Majewski, J., 2010. Estimated Discard and Total Catch of Selected Groundfish Species in the 2008 U.S. West Coast Fisheries. West Coast Groundfish Observer Program.

Bernatis, J., Gerstenberger, S., McGaw, I., 2007. Behavioural responses of the Dungeness crab, *Cancer magister*, during feeding and digestion in hypoxic conditions. Mar. Biol. 150, 941–951.

Bodkin, J., Esslinger, G., Monson, D., 2004. Foraging depths of sea otters and implications to coastal marine communities. Mar. Mamm. Sci. 20, 305–321.

Booth, J., Phillips, A., Jamieson, G., 1986. Fine scale spatial distribution of *Cancer magister* megalopae and its relevance to sampling methodology. In: Melteff, B. (Ed.), Proceedings of the Symposium on Dungeness Crab Biology and Management. University of Alaska Sea Grant, Anchorage, Alaska, pp. 273–286.

Botsford, L., 1981. Comment on cycles in the northern California Dungeness crab population. Can. J. Fish. Aquat. Sci. 38, 1295–1296.

Botsford, L., 1984. Effect of individual growth rates on expected behavior of the Northern California Dungeness crab (*Cancer magister*) fishery. Can. J. Fish. Aquat. Sci. 41, 99–107.

Botsford, L., Hobbs, R., 1995. Recent advances in the understanding of cyclic behavior of Dungeness crab (*Cancer magister*) populations. ICES J. Mar. Sci. 199, 157–166.

Botsford, L., Lawrence, C., 2002. Patters of co-variability among California Current chinook salmon, coho salmon, Dungeness crab, and physical oceanographic conditions. Prog Oceangr. 53, 283–305.

Botsford, L., Methot, R., Wilen, J., 1982. Cyclic covariation in the California king salmon, *Oncorhynchus tshawytscha*, silver salmon, *O. kisutch*, and Dungeness crab, *Cancer magister*, fisheries. Fish B-NOAA 80, 791–801.

Botsford, L., Wickham, D., 1978. Behavior of age specific, density-dependent models and the northern California Dungeness crab (*Cancer magister*) fishery. J. Fish. Res. Board Can. 35, 833–843.

Botsford, L.W., Methot, R., Johnston, W., 1983. Effort dynamics of the Northern California Dungeness crab (*Cancer magister*) fishery. Can. J. Fish. Aquat. Sci. 40, 337–346.

Botsford, L.W., Wickham, D., 1975. Correlation of upwelling index and Dungeness crab catch. Fish B-NOAA 73, 901–907.

Breen, P., 1985a. Ghost fishing by Dungeness crab traps: a preliminary report. Can. Manuscr. Rep. Fish. Aquat. Sci. 1848, 51–55.

Breen, P., 1985b. Crab gear selectivity studies in Departure Bay. Can. Manuscr. Rep. Fish. Aquat. Sci. 1848, 21–39.

Buchanan, D., 1970. Effects of the insecticide Sevin on various stages of the Dungeness crab, *Cancer magister*. J. Fish. Res. Board Can. 27, 93–104.

Buchanan, D., Milleman, R., 1969. The prezoeal stage of the Dungeness crab, *Cancer magister* Dana. Biol. Bull. 137, 250–255.

Burgess, A., 2011. Vectoring Algal Toxin in Marine Planktonic Food Webs: Sorting Out Nutritional Deficiency from Toxicity Effects. MS, Western Washington University.

Butler, T., 1954. Food of the commercial crab in the Queen Charlotte Islands region. Can. Fish. Res. Bd. Pac. Prog. Rep. 99, 3–5.

Butler, T., 1960. Maturity and breeding of the Pacific edible crab, *Cancer magister* Dana. J. Fish. Res. Board Can. 17, 641–646.

Butler, T., 1961. Growth and age determination of the Pacific edible crab, *Cancer magister* Dana. J. Fish. Res. Board Can. 18, 873–891.

Butler, T., Hankin, D., 1992. Comment on mortality rate of Dungeness crabs (*Cancer magister*). Can. J. Fish. Aquat. Sci. 49, 1518–1525.

Caldwell, R., Armstrong, D., Buchanan, D., Mallon, M., Milleman, R., 1978. Toxicity of the fungicide captan to the Dungeness crab, *Cancer magister*. Mar. Biol. 48, 11–17.

Caldwell, R., Buchanan, D., Armstrong, D., Mallon, M., Millemann, R., 1979. Toxicity of the herbicides 2,4-D, DEF, propanil and trifluralin to the Dungeness crab, *Cancer magister*. Arch. Environ. Contam. Toxicol. 8, 383–396.

Canada. Dept. of Fish. and Oceans. Pacific Region. 2012. Pacific region integrated fisheries management plan: crab by trap. 220 p.

Carrasco, K.R., Armstrong, D., Gunderson, D., Rogers, C., 1985. Abundance and growth of *Cancer magister* young-of-the-year in the nearshore environment. In: Melteff, B. (Ed.), Proceedings of the Symposium on Dungeness Crab Biology and Management. University of Alaska Sea Grant, Anchorage, Alaska, pp. 171–184.

Chan, F., Barth, J., Lubchenco, J., Kirincich, A., Weeks, H., Peterson, W., Menge, B., 2008. Emergence of anoxia in the California Current large marine ecosystem. Science 319, 920.

Chang, B., Levings, C., 1978. Effects of burial on the heart cockle *Clinocardium nuttallii* and the Dungeness crab *Cancer magister*. Estuar. Coast. Shelf Sci. 7, 409–412.

Childers, R., Reno, P., Olson, R., 1996. Prevalence and geographic range of *Nadelspora canceri* (Microspora) in Dungeness crab *Cancer magister*. Dis. Aquat. Organ. 24, 135–142.

Cleaver, F., 1949. Preliminary results of the coastal crab (*Cancer magister*) investigation. Wash. Dep. Fish. 49, 47–82.

Coleman, R.A., Hoskin, M.G., Von Carlshausen, E., Davis, C.M., 2013. Using a no-take zone to assess the impacts of fishing: sessile epifauna appear insensitive to environmental disturbances from commercial potting. J. Exp. Mar. Biol. Ecol. 440, 100–107.

Collier, P., 1983. Movement and growth of post-larval Dungeness crabs, *Cancer magister*, in the San Francisco area. Fish Bull. (Calif.) 172, 125–134.

Collins, C., Garfield, N., Rago, T., Rischmiller, F., Carter, E., 2000. Mean structure of the inshore countercurrent and California undercurrent off Point Sur, California. Deep Sea Res. Part I 47, 765–782.

Cumbrow, R., 1978. Washington State Shellfish Information Booklet. Washington Department of Fisheries, 55 pp.

Curtis, D., McGaw, I., 2008. A year in the life of a Dungeness crab: methodology for determining microhabitat conditions experienced by large decapod crustaceans in estuaries. Trans. Zool. Soc. London 274, 375–385.

Curtis, D., Vanier, C., McGaw, I., 2010. The effects of starvation and acute low salinity exposure on food intake in the Dungeness crab, *Cancer magister*. Mar. Biol. 157, 603–612.

Dahlstrom, W., Wild, P., 1983. A history of Dungeness crab fisheries in California. Fish Bull. (Calif.) 172, 7–23.

Dana, J., 1852. Conspectus crustaceorum, conspectus of the crustacea of the exploring expedition under Capt. Wilkes, U.S.N. including the Crustacea Cancroidea and Corystoidea. Proc. Acad. Nat. Sci. Phila. 6, 72–105.

Demory, D., 1990. History and status of the Oregon Dungeness crab fishery. Oregon Department of Fish and Wildlife, Salem, OR, 12 p.

Dewees, C., Sortais, K., Krachey, M., Hackett, S., Hankin, D., 2004. Racing for crab... Costs and management options evaluated in Dungeness crab fishery. Calif. Agr. 58, 186–193.

Diamond, N., Hankin, D., 1985. Movements of adult female Dungeness crabs (*Cancer magister*) in northern California based on tag recoveries. Can. J. Fish. Aquat. Sci. 42, 919–926.

Didier, A.J., 2002. The Pacific coast Dungeness crab fishery. Gladstone, OR: Pacific states marines fisheries commission. 30 p.

Dinnel, P., Armstrong, D., Mcmillan, R., 1993. Evidence for multiple recruitment-cohorts of Puget Sound Dungeness crab, *Cancer magister*. Mar. Biol. 115, 53–63.

Dumbauld, B., 1993. Use of oyster shell to enhance intertidal habitat and mitigate loss of Dungeness crab (*Cancer magister*) caused by dredging. Can. J. Fish. Aquat. Sci. 50, 381–390.

Dunn, P., 2011. Larval Biology and Estuarine Ecology of the Nemertean Egg Predator *Carcinonemertes errans* on the Dungeness crab, *Cancer magister*. Doctor of Philosophy, University of Oregon.

Dunn, P., Shanks, A., 2012. Mating success of female Dungeness crabs (*Cancer magister*) in Oregon coastal waters. J. Shellfish Res. 31, 835–839.

Ebert, E., Haseltine, A., Houk, J., Kelly, R., 1983. Laboratory cultivation of the Dungeness crab, *Cancer magister*. Fish Bull. (Calif.) 172, 259–310.

Eggleston, D., Armstrong, D., 1995. Presettlement and postsettlement determinants of estuarine Dungeness crab recruitment. Ecol. Monogr. 65, 193–216.

Eggleston, D., Armstrong, D., Elis, W., Patton, W., 1998. Estuarine fronts as conduits for larval transport: hydrodynamics and spatial distribution of Dungeness crab postlarvae. Mar. Ecol. Prog. Ser. 164, 73–82.

Emmett, R., Krutzikowsky, G., 2008. Nocturnal feeding of Pacific hake and jack mackerel off the mouth of the Columbia River, 1998-2004: implications for juvenile salmon predation. Trans. Am. Fish. Soc. 137, 657–676.

Esri, 2011. ARCGIS Desktp: Release 10. Environmental Systems Research Institute, Redlands, CA.

Favorite, F., 1967. The Alaskan Stream. Int. North Pac. Fish. Comm. Bull. 21, 1–20.

Fernandez, M., 1999. Cannibalism in Dungeness crab *Cancer magister*: effects of predator-prey size ratio, density, and habitat type. Mar. Ecol. Prog. Ser. 182, 221–230.

Fernandez, M., Armstrong, D., Iribarne, O., 1993. First cohort of young-of-the-year Dungeness crab, *Cancer magister*, reduces abundance of subsequent cohorts in intertidal shell habitat. Can. J. Fish. Aquat. Sci. 50, 2100–2105.

Fernandez, M., Iribarne, O., Armstrong, D., 1994a. Ecdysial rhythms in megalopae and 1st instars of the Dungeness crab, *Cancer magister*. Mar. Biol. 118, 611–615.

Fernandez, M., Iribarne, O., Armstrong, D., 1994b. Swimming behavior of Dungeness crab, *Cancer magister*, megalopae in still and moving water. Estuaries 17, 271–275.

Fisher, J., 2006. Seasonal timing and duration of brachyuran larvae in a high-latitude fjord. Mar. Ecol. Prog. Ser. 323, 213–222.

Fisher, W., Nelson, R., 1977. Therapeutic treatment for epibiotic fouling on Dungeness crab (*Cancer magister*) larvae reared in the laboratory. J. Fish. Res. Board Can. 34, 432–436.

Fisher, W., Nelson, R., 1978. Application of antibiotics in the cultivation of Dungeness crab, *Cancer magister*. J. Fish. Res. Board Can. 35, 1343–1349.

Fisher, W., Velasquez, D., 2008. Management Recommendations for Washington's Priority Habitats and Species Dungeness Crab *Cancer magister*. Washington Department of Fisheries, Olympia, WA.

Fisher, W., Wickham, D., 1976. Mortalities and epibiotic fouling of eggs from wild populations of Dungeness crab, *Cancer magister*. Fish B-NOAA 74, 201–207.

Froehlich, H., Essington, T., Beaudreau, A., Levin, P., in review. Movement patterns and distributional shifts of Dungeness crabs (*Metacarcinus magister*) and English sole (*Parophrys vetulus*) during seasonal hypoxia. Estuar Coast Shelf S.

Garcia, R., Sukin, S., Lopez, M., 2011. Effects on larval crabs of exposure to algal toxin via ingestion of heterotrophic prey. Mar. Biol. 158, 451–460.

Garshelis, D., Garshelis, J., Kimker, A., 1986. Sea otter time budgets and prey relationships in Alaska. J. Wildl. Manage. 50, 637–647.

Gaumer, T., 1969. Annual report: controlled rearing of Dungeness crab larvae and the influence of environmental conditions on their survival. US Department of Interior Fish and Wildlife Service, Contract 14-17-0001-1907.

Gaumer, T., 1970. Annual report: controlled rearing of Dungeness crab larvae and the influence of environmental conditions on their survival. US Department of Interior Fish and Wildlife Service, Contract 14-17-0001-2131.

Gaumer, T., 1971. Closing report: controlled rearing of Dungeness crab larvae and the influence of environmental conditions on their survival. US Department of Interior Fish and Wildlife Service, Contract 14-17-0001-2325.

Gelchu, A., Pauly, D., 2007. Growth and distribution of port-based global fishing effort within countries' EEZs from 1970 to 1995. Fish. Centre Res. Rep. 15, 1–99.

Gotshall, D., 1977. Stomach contents of northern California Dungeness crabs, *Cancer magister*. Fish Bull. (Calif.) 63, 43–51.

Greengo, R., 1952. Shellfish foods of the California Indians. Kroeber Anthropol. Soc. 7, 63–114.

Gunderson, D., Armstrong, D., Shi, Y., McConnaughey, R., 1990. Patterns of estuarine use by juvenile English Sole (*Parophrys vetulus*) and Dungeness crab (*Cancer magister*). Estuaries 13, 59–71.

Gunderson, D., Ellis, I., 1986. Development of a plumb staff beam trawl for sampling demersal fauna. Fish. Res. 4, 35–41.

Gutermuth, F., 1989. Temperature-dependent metabolic response of juvenile Dungeness crab *Cancer magister* Dana: ecological implications for estuarine and coastal populations. J. Exp. Mar. Biol. Ecol. 126, 135–144.

Hackett, S., Krachey, M., Dewees, C., Hankin, D., Sortais, K., 2003. An economic overview of Dungeness crab (*Cancer magister*) processing in California. CalCOFI Rep. 44, 86–93.

Hankin, D., 1985. Proposed explanations for fluctuations in abundance of Dungeness crabs: a review and critique. In: Melteff, B. (Ed.), Proceedings of the Symposium on Dungeness Crab Biology and Managment. University of Alaska Sea Grant, Fairbanks.

Hankin, D., Butler, T., Wild, P., Xue, Q., 1997. Does intense fishing on males impair mating success of female Dungeness crabs? Can. J. Fish. Aquat. Sci. 54, 655–669.

Hankin, D., Diamond, N., Mohr, M., Ianelli, J., 1989. Growth of reproductive dynamics of adult female Dungeness crabs (*Cancer magister*) in Northern California. J. Conseil 46, 94–108.

Harrison, M., Crespi, B.J., 1999. Phylogenetics of Cancer crabs (Crustacea: Decapoda: Brachyura). Mol. Phylogenet. Evol. 12, 186–199.

Hartman, M. 1977. A Mass Rearing System for the Culture of Brachyuran crab Larvae. Proceedings of the 8th Annual Workshop of the World Mariculture Society, vol. 8, pp. 147–155.

Hartman, M., Letterman, G., 1978. An evaluation of three species of diatoms as food for *Cancer magister* larvae. World Maricult. Soc. 9, 271–276.

Hartnoll, R., 1969. Mating in the brachyura. Crustaceana 16, 161–181.

Hatfield, S., 1983. Intermoult staging and distribution of Dungeness crab, *Cancer magister*, megalopae. Fish Bull. (Calif.) 172, 85–96.

Herter, H., Eckert, G.L., 2008. Transport of Dungeness crab, *Cancer magister*, megalopae into Glacier Bay, Alaska. Mar. Ecol. Prog. Ser. 372, 181–194.

Hickey, B., 1979. The California current system: hypotheses and facts. Prog. Oceanogr. 8, 191–279.

Higgins, K., Hastings, A., Sarvela, J., Botsford, L.W., 1997. Stochastic dynamics and deterministic skeletons: population behavior of Dungeness crab. Science 276, 1431–1435.

High, W., 1976. Escape of Dungeness crabs from pots. Mar. Fish. Rev. 38, 19–23.

Hildenbrand, K., Gladics, A., Eder, B., 2011. Adult male Dungeness crab (*Metacarcinus magister*) movements near Reedsport Oregon from a fisheries collaborative mark-recapture study. Oregon Wave Energy Trust and the Oregon Dungeness Crab Commission.

Hobbs, R., Botsford, L.W., 1992. Diel vertical migration and timing of metamorphosis of larvae of the Dungeness crab *Cancer magister*. Mar. Biol. 112, 417–428.

Hobbs, R., Botsford, L.W., Thomas, A., 1992. Influence of hydrographic conditions and wind forcing on the distribution and abundance of Dungeness crab, *Cancer magister*, larvae. Can. J. Fish. Aquat. Sci. 49, 1379–1388.

Holsman, K., Armstrong, D.A., Beauchamp, D., Ruesink, J., 2003. The necessity for intertidal foraging by estuarine populations of subadult Dungeness crab, *Cancer magister*: evidence from a bioenergetics model. Estuaries 26, 1155–1173.

Holsman, K., Mcdonald, P., Armstrong, D.A., 2006. Intertidal migration and habitat use by subadult Dungeness crab *Cancer magister* in a NE Pacific estuary. Mar. Ecol. Prog. Ser. 308, 183–195.

Hooff, R.C., Peterson, W.T., 2006. Copepod biodiversity as an indicator of changes in ocean and climate conditions of the northern California current ecosystem. Limnol. Oceanogr. 51, 2607–2620.

Huyer, A., 1977. Seasonal variation in temperature, salinity, and density over the continental shelf off Oregon. Limnol. Oceanogr. 22, 442–453.

Huyer, A., 1983. Coastal upwelling in the California current system. Prog. Oceanogr. 12, 259–284.

Huyer, A., Kosro, P.M., Lentz, S.J., Beardsley, R., 1989. Poleward flow in the California Current system. In: Neshyba, S.J., Mooers, C.N.K., Smith, R.L., Barber, R.T.

(Eds.), Poleward Flows along Eastern Ocean Boundaries, Coastal and Estuarine Studies, vol. 34. Springer, Berlin, pp. 144–159.

Huyer, A., Sobey, E., Smith, R., 1979. The spring transition in currents over the Oregon continental shelf. J. Geophys. Res.-Oc. Atm. 84, 6995–7011.

Iribarne, O., Armstrong, D., Fernandez, M., 1995. Environmental-impact of intertidal juvenile Dungeness crab habitat enhancement—effects on bivalves and crab foraging rate. J. Exp. Mar. Biol. Ecol. 192, 173–194.

Iribarne, O., Fernandez, M., Armstrong, D., 1994. Does space competition regulate density of juvenile Dungeness crab *Cancer magister* Dana in sheltered habitats? J. Exp. Mar. Biol. Ecol. 183, 259–271.

Jacoby, C., 1982. Behavioral responses of the larvae of *Cancer magister* Dana (1852) to light, pressure and gravity. Mar. Behav. Physiol. 8, 267–283.

Jacoby, C., 1983. Ontogeny of behavior in the crab instars of the Dungeness crab, *Cancer magister* Dana 1852. Z. Tierpsychol. 63, 1–16.

Jaffe, L., Nyblade, C., Forward, R., Sulkin, S., 1987. Phylum or Subphylum Crustacea, class Malacostraca, Order Decapoda, Brachyura. In: Strathmann, M. (Ed.), Reproductive and Development of Marine Invertebrates of the Northern Pacific Coast: Data and Methods for the Study of Eggs, Embryos and Larvae. University of Washington Press, Seattle.

Jamieson, G., Phillips, A., 1988. Occurrence of Cancer crab (*C. magister* and *C. oregonensis*) megalopae off the west coast of Vancouver Island, British Columbia. Fish B-NOAA 86, 525–542.

Jamieson, G., Phillips, A., 1993. Megalopal spatial-distribution and stock separation in Dungeness crab (*Cancer magister*). Can. J. Fish. Aquat. Sci. 50, 416–429.

Jamieson, G., Phillips, A., Huggett, W., 1989. Effects of ocean variability on the abundance of Dungeness crab (*Cancer magister*) megalopae. Can. J. Fish. Aquat. Sci. 108, 305–325.

Jensen, G., 1995. Pacific Coast Crabs and Shrimps. Sea Challengers, Monterey, CA, 87 pp.

Jensen, G., Armstrong, D., 1987. Range extensions of some Northeastern Pacific decapoda. Crustaceana 52, 215–217.

Jensen, G., Asplen, M., 1998. Omnivory in the diet of juvenile dungeness crab, *Cancer magister* Dana. J. Exp. Mar. Biol. Ecol. 226, 175–182.

Jensen, P.C., Bentzen, P., 2012. A molecular dissection of the mating system of the Dungeness crab, *Metacarcinus magister* (Brachyura: Cancridae). J. Crustacean Biol. 32, 443–456.

Jensen, P.C., Orensanz, J., Armstrong, D.A., 1996. Structure of the female reproductive tract in the Dungeness crab (*Cancer magister*) and implications for the mating system. Biol. Bull. 190, 336–349.

Johnson, D., Botsford, L.W., Methot, R., Wainwright, T., 1986. Wind stress and cycles in Dungeness crab (*Cancer magister*) catch off California, Oregon and Washington. Can. J. Fish. Aquat. Sci. 43, 838–845.

Johnson, J., Shanks, A.L., 2002. Time series of the abundance of the post-larvae of the crabs *Cancer magister* and Cancer spp. on the southern Oregon coast and their cross-shelf transport. Estuaries 25, 1138–1142.

Juanes, F., Hartwick, E., 1990. Prey size selection in Dungeness crabs: the effect of claw damage. Ecology 71, 744–758.

Keister, J., Di Lorenzo, E., Morgan, C., Combes, V., Peterson, W., 2011. Zooplankton species composition is linked to ocean transport in the Northern California Current. Glob. Change Biol. 17, 2498–2511.

Kline, T., 2002. The relative trophic position of *Cancer magister* megalopae within the planktonic community of the sub-polar northeastern Pacific Ocean. In: Paul, A., Dawe, E., Elner, R., Jamieson, G., Kruse, G.H., Otto, R., Sainte-Marie, B., Shirley, T., Woodby, D. (Eds.), Crabs in Cold Water Regions: Biology, Management and Economics. University of Alaska Sea Grant, Fairbanks, pp. 645–649.

Knudsen, J., 1964. Observations of the reproductive cycles and ecology of the common Brachyura and crablike Anomura of Puget Sound, Washington. Pac. Sci. 18, 3–33.

Kuris, A., Sadeghian, P., Carlton, J., 2007. Keys to decapod Crustacea. In: Carlton, J. (Ed.), The Light and Smith Manual: Intertidal Invertebrates from Central California to Oregon. University of California Press, Berkeley, pp. 631–656.

Lawton, P., Elner, R., 1985. Feeding in relation to morphometrics within the genus *Cancer*: evolutionary and ecological considerations. In: Melteff, B. (Ed.), Proceedings of the Symposium on Dungeness Crab Biology and Management. University of Alaska Sea Grant, Fairbanks, pp. 357–380.

Lewis, C., Slade, S., Maxwell, K., Mathews, T., 2009. Lobster trap impact on coral reefs: effects of wind-driven trap movement. N. Z. J. Mar. Freshw. Res. 43, 271–282.

Losey, R.J., Yamada, S B., Largaespada, L., 2004. Late-holocene Dungeness crab (*Cancer magister*) harvest at an Oregon coast estuary. J. Arachaeol. Sci. 31, 1603–1612.

Lough, R., 1975. Dynamics of Crab Larvae (Anomura, Brachyura) off the Central Oregon Coast, 1969–1971, Oregon State University.

Lough, R., 1976. Larval dynamics of the Dungeness crab, *Cancer magister*, off the central Oregon coast, 1970-71. Fish B-NOAA 74, 353–376.

Love, M., Westphal, W., 1981. A correlation between annual catches of Dungeness crab, *Cancer magister*, and mean annual sunspot number. Fish B-NOAA 79, 794–795.

Lynn, R., Simpson, J., 1987. The California current system: the seasonal variability of its physical characteristics. J. Geophys. Res. Oceans 92, 12947–12966.

Mackay, D., 1942. The pacific edible crab, *Cancer magister*. Fish. Res. Bd. Can. 62, 255–265.

Mann, K., Lazier, J., 2006. Dynamics of Marine Ecosystems. Wiley-Blackwell, Malden, MA.

Mantua, N., Hare, S., 2002. The Pacific decadal oscillation. J. Oceanogr. 58, 35–44.

Mantua, N., Hare, S., Zhang, Y., Wallace, J., Francis, R., 1997. A Pacific interdecadal climate oscillation with impacts on salmon production. Bull. Am. Meteor. Soc. 78, 1069–1079.

Martin, M., Osborn, K., Billig, P., Glickstein, N., 1981. Toxicities of ten metals to *Crassostrea gigas* and *Mytilus edulis* embryos and *Cancer magister* larvae. Mar. Pollut. Bull. 12, 305–308.

Mayer, D., 1973. The Ecology and Thermal Sensitivity of the Dungeness crab, *Cancer magister*, and Related Species of Its Benthic Community on Similk Bay, Washington. PhD, University of Washington.

McConnaughey, R., Armstrong, D., Hickey, B., 1995. Dungeness crab (*Cancer magister*) recruitment variability and Ekman transport of larvae. ICES J. Mar. Sci. 199, 167–174.

McConnaughey, R., Armstrong, D., Hickey, B., Gunderson, D., 1992. Juvenile Dungeness crab (*Cancer magister*) recruitment variability and oceanic transport during the pelagic larval phase. Can. J. Fish. Aquat. Sci. 49, 2028–2044.

McConnaughey, R., Armstrong, D., Hickey, B., Gunderson, D., 1994. Interannual variability in coastal Washington Dungeness crab (*Cancer magister*) populations: larval advection and the coastal landing strip. Fish. Oceanogr. 3, 22–38.

McConnaughey, R.A., Armstrong, D.A., 1995. Potential effects of global climate change on Dungeness crab (Cancer magister) populations in the northeastern Pacific Ocean. In: Beamish, R.J. (Ed.), Climate Change and Northern Fish Populations. Canadian Special Publication of Fisheries and Aquatic Sciences. vol. 121. National Research Council Monograph Publishing Program, Ontario, Canada, pp. 291–306.

McConnaughey, R.A., Conquest, L., 1993. Trawl survey estimation using a comparative approach based on lognormal theory. Fish B-NOAA 91, 107–118.

McDonald, P., Jensen, G., Armstrong, D., 2001. The competitive and predatory impacts of the nonindigenous crab *Carcinus maenas* (L.) on early benthic phase Dungeness crab *Cancer magister* Dana. J. Exp. Mar. Biol. Ecol. 258, 39–54.

McGaw, I., 2005. Burying behaviour of two sympatric crab species: *Cancer magister* and *Cancer productus*. Sci. Mar. 69, 375–381.

McGaw, I., 2008. Gastric processing in the Dungeness crab, *Cancer magister*, during hypoxia. Comp. Biochem. Physiol. 150, 458–463.

McKelvey, R., Hankin, D., 1981. Reply to comment on cycles in the northern California Dungeness crab population. Can. J. Fish. Aquat. Sci. 38, 1296–1297.

McKelvey, R., Hankin, D., Yanosko, K., Snygg, C., 1980. Stable cycles in multistage recruitment models—an application to the Northern California Dungeness crab (*Cancer magister*) fishery. Can. J. Fish. Aquat. Sci. 37, 2323–2345.

McMillan, R., Armstrong, D., Dinnel, P., 1995. Comparison of intertidal habitat use and growth rates of two northern Puget Sound cohorts of 0+ age Dungeness crab, *Cancer magister*. Estuaries 18, 390–398.

Melteff, B., 1985. Proceedings of the Symposium on Dungeness Crab Biology and Management, Anchorage, Aaska, University of Alaska Sea Grant.

Methot, R., 1989. Management of a cyclic resource: the Dungeness crab fisheries of the Pacific Coast of North America. In: Caddy, J. (Ed.), Marine Invertebrate Fisheries: Their Assessment and Management. John Wiley & Sons, United States of America, pp. 205–223.

Methot, R., Botsford, L.W., 1982. Estimated preseason abundance in the California Dungeness crab (*Cancer magister*) fisheries. Can. J. Fish. Aquat. Sci. 39, 1077–1083.

Miller, J., Shanks, A., 2004. Ocean-estuary coupling in the Oregon upwelling region: abundance and transport of juvenile fish and of crab megalopae. Mar. Ecol. Prog. Ser. 271, 267–279.

Miller, T., Hankin, D., 2004. Descriptions and durations of premolt setal stages in female Dungeness crabs, *Cancer magister*. Mar. Biol. 144, 101–110.

Minobe, S., Mantua, N., 1999. Interdecadal modulation of interannual atmospheric and oceanic variability over the North Pacific. Prog. Oceanogr. 43, 163–192.

Mohr, M., Hankin, D., 1989. Estimation of size-specific molting probabilities in adult decapod crustaceans based on postmolt indicator data. Can. J. Fish. Aquat. Sci. 46, 1819–1830.

Moloney, C., Botsford, L., Largier, J., 1994. Development, survival and timing of metamorphosis of planktonic larve in a variable environment—the Dungeness crab as an example. Mar. Ecol. Prog. Ser. 113, 61–79.

Morado, J., Sparks, A., 1988. A review of infectious diseases of the Dungeness crab, *Cancer magister*. J. Shellfish Res. 7, 127.

Morgan, S.G., 1995. The timing of larval release. In: McEdward, L. (Ed.), Ecology of Marine Invertebrate Larvae. CRC Press, Boca Raton, FL, pp. 157–192.

Northrup, T., 1975. Completion report: coastal Dungeness crab study. National Oceanic and Atmospheric Association.

Oh, S., Hankin, D., 2004. The sperm plug is a reliable indicator of mating success in female Dungeness crabs, *Cancer magister*. J. Crustacean Biol. 24, 314–326.

Orensanz, J., Armstrong, J., Armstrong, D.A., Hilborn, R., 1998. Crustacean resources are vulnerable to serial depletion—the multifaceted decline of crab and shrimp fisheries in the Greater Gulf of Alaska. Rev. Fish Biol. Fisher. 8, 117–176.

Orensanz, J., Gallucci, V., 1988. Comparative study of postlarval life-history schedules in four sympatric species of cancer (Decapoda: Brachyura: Cancridae). J. Crustacean Biol. 8, 187–220.

Pane, E., Barry, J., 2007. Extracellular acid-base regulation during short-term hypercapnia is effective in a shallow-water crab, but ineffective in a deep-sea crab. Mar. Ecol. Prog. Ser. 334, 1–9.

Park, W., Douglas, D., Shirley, T., 2007. North to Alaska: evidence for conveyor belt transport of Dungeness crab larvae along the west coast of the United States and Canada. Limnol. Oceanogr. 52, 248–256.

Park, W., Shirley, T., 2005. Diel vertical migration and seasonal timing of the larvae of three sympatric cancrid crabs, *Cancer* spp., in southeastern Alaska. Estuaries 28, 266–273.

Park, W., Shirley, T., 2008. Development and distribution of Dungeness crab larvae in Glacier Bay and neighboring straits in Southeastern Alaska: implications for larval advection and retention. Anim. Cells Syst. 12, 279–286.

Pauley, G., Armstrong, D., Heun, T., 1986. Species profiles: life histories and environmental requirements of coastal fishes and invertebrates (Pacific Northwest), Dungeness crab. Biol. Rep. 82, 20.

Pearson, W., 1979. Thresholds for detection and feeding behavior in the Dungeness crab, Cancer magister (Dana). J. Exp. Mar. Biol. Ecol. 39, 65–78.

Pearson, W., 1981. Effects of oiled sediment on predation on the littleneck clam, Protothaca staminea, by the Dungeness crab, Cancer magister. Estuar. Coast. Shelf Sci. 13, 445–454.

Peterson, W., 1973. Upwelling indices and annual catches of Dungeness crab, Cancer magister, along the West coast of the United States. Fish B-NOAA 71, 902–910.

Pierce, S., Smith, R., Kosro, P.M., Barth, J.A., Wilson, C., 2000. Continuity of the poleward undercurrent along the eastern boundary of the mid-latitude north Pacific. Deep-Sea Res. Part II 47, 811–829.

Poole, R., 1966. A description of laboratory-reared zoeae of Cancer magister Dana, and megalopae taken under natural conditions (Decapoda Brachyura). Crustaceana 11, 83–97.

Reed, P., 1966. Annual report: controlled rearing of Dungeness crab larvae and the influence of environmental conditions on their survival. US Department of Interior Fish and Wildlife Service, Contract 14-17-0007-353.

Reed, P., 1969. Culture methods and effects of temperature and salinity on survival and growth of Dungeness crab (Cancer magister) larvae in the laboratory. J. Fish. Res. Board Can. 26, 389–397.

Reed, R., Halpern, D., 1976. Observations of the California Undercurrent off Washington and Vancouver Island. Limnol. Oceanogr. 21, 389–398.

Reilly, P., 1983a. Dynamics of Dungeness crab, Cancer magister, larvae off central and northern California. Fish Bull. (Calif.) 172, 57–84.

Reilly, P., 1983b. Effects of commercial trawling on Dungeness crab survival. Fish Bull. (Calif.) 172, 165–174.

Reilly, P., 1983c. Predation on Dungeness crab, Cancer magister, in central California. Fish Bull. (Calif.) 172, 155–164.

Roegner, G., Armstrong, D., Hickey, B., Shanks, A., 2003. Ocean distribution of Dungeness crab megalopae and recruitment patterns to estuaries in southern Washington State. Estuaries 26, 1058–1070.

Roegner, G.C., Armstrong, D.A., Shanks, A.L., 2007. Wind and tidal influences on larval crab recruitment to an Oregon estuary. Mar. Ecol. Prog. Ser. 351, 177–188.

Romsos, C., 2004. Mapping Surficial Geologic Habitats of the Oregon Continental Margin Using Integrated Interpretive GIS Techniques. MS, Oregon State University.

Ruppert, E., 1994. Invertebrate Zoology: A Functional Evolutionary Approach. Australia, Thomson Brooks/Cole.

Ruttimann, J., 2006. Oceanography: sick seas. Nature 442, 978.

Scheding, K., Shirley, T., O'clair, C., Taggart, S., 1999. Critical habitat for ovigerous Dungeness crabs. In: Kruse, G.H., Bez, N., Booth, A., Dorn, M., Hills, S., Lipcius, R., Pelletier, D., Roy, C., Smith, S., Witherell, D. (Eds.), Spatial Processes and Management of Marine Populations. University of Alaska Sea Grant, Anchorage.

Schweitzer, C., Feldmann, R., 2000. Re-evaluation of the Cancridae Latreille, 1802 (Decapoda: Brachyura) including three new genera and three new species. Bijdr Dierkd 69, 223–250.

Shanks, A., in press. Atmospheric forcing drives recruitment variation in the Dungeness crab (Cancer magister), revisited. Fish Oceanogr.

Shanks, A.L., 2006. Mechanisms of cross-shelf transport of crab megalopae inferred from a time series of daily abundance. Mar. Biol. 148, 1383–1398.

Shanks, A.L., Roegner, G.C., 2007. Recruitment limitation in Dungeness crab populations is driven by variation in atmospheric forcing. Ecology 88, 1726–1737.

Shanks, A.L., Roegner, G.C., Miller, J., 2010. Using megalopae abundance to predict future commercial catches of Dungeness crab (Cancer magister) in Oregon. CalCOFI report, 51 pp.

Shenker, J., 1988. Oceanographic associations of neustonic larval and juvenile fishes and Dungeness crab megalopae off Oregon. Fish B-NOAA 86, 299–317.

Shirley, S., Shirley, T., 1988. Appendage injury in Dungeness crabs in Southeastern Alaska. Fish B-NOAA 86, 156–160.

Shirley, S., Shirley, T., Rice, S., 1987. Latitudinal variation in the Dungeness crab, Cancer magister: zoeal morphology explained by incubation temperature. Mar. Biol. 95, 371–376.

Shirley, T., Bishop, G., O'clair, C., Taggart, S., Bodkin, J., 1996. Sea Otter predation on Dungeness crabs in Glacier bay, Alaska. In: High Latitude Crabs: Biology, Management and Economics. University of Alaska Sea Grant, Anchorage Alaska.

Sigler, M., Cameron, M., Eagleton, M., Faunce, C., Heifetz, J., Helser, T., Laurel, B., Lindberg, M., McConnaughey, R., Ryer, C., Wilderbeur, T. 2012. Alaska essential fish habitat research plan: a research plan for the National Marine Fisheries Service's Alaska Fisheries Science Center and Alaska regional office. AFSC processed report, 21 pp.

Smith, B., Jamieson, G., 1989a. Exploitation and mortality of male Dungeness crabs (Cancer magister) near Tofino, British Columbia. Can. J. Fish. Aquat. Sci. 46, 1609–1614.

Smith, B., Jamieson, G., 1989b. A model for standardizing Dungeness crab (Cancer magister) catch rates among traps which experienced different soak times. Can. J. Fish. Aquat. Sci. 46, 1600–1608.

Smith, B., Jamieson, G., 1991a. Movement, spatial-distribution, and mortality of male and female Dungeness crab Cancer magister near Tofino, British Columbia. Fish B-NOAA 89, 137–148.

Smith, B., Jamieson, G., 1991b. Possible consequences of intensive fishing for males on the mating opportunities of Dungeness crabs. Trans. Am. Fish. Soc. 120, 650–653.

Smith, B., Jamieson, G., 1992. Reply to Butler and Hankin: mortality rates of Dungeness crabs (Cancer magister). Can. J. Fish. Aquat. Sci. 49, 1521–1525.

Smith, Q., Eckert, G., 2011. Spatial variation and evidence for multiple transport pathways for Dungeness crab Cancer magister late-stage larvae in southeastern Alaska. Mar. Ecol. Prog. Ser. 429, 185–196.

Snow, C., Wagner, E., 1965. Tagging of Dungeness crabs with spaghetti and dart tags. Fish Commun. Oreg. 4629, 5–13.

Snow, S., Nielson, J., 1966. Premating and mating behavior of Dungeness crab (Cancer magister Dana). J. Fish. Res. Board Can. 23, 1319–1323.

Sparks, A., Morado, J., Hawkes, J., 1985. A systemic microbial disease in the Dungeness crab, Cancer magister, caused by a Chlamydia-like organism. J. Invertebr. Pathol. 45, 204–217.

Spencer, M., Stoner, A., Ryer, C., Munk, J., 2005. A towed camera sled for estimating abundance of juvenile flatfishes and habitat characteristics: comparison with beam trawls and divers. Estuar. Coast. Shelf Sci. 64, 497–503.

Stabeno, P., Bond, N., Hermann, A., Kachel, N., Mordy, C., Overland, J., 2004. Meteorology and oceanography of the Northern Gulf of Alaska. Cont. Shelf Res. 24, 859–897.

Stefferud, J.A., 1975. Prediction of Abundance of Harvestable Dungeness crab (Cancer magister). MS, Oregon State University.

Stevens, B., 2003. Timing of aggregation and larval release by Tanner crabs, Chionoecetes bairdi, in relation to tidal current patterns. Fish. Res. 65, 201–216.

Stevens, B., 2006. Timing and duration of larval hatching for blue king crab *Paralithodes platypus* Brandt, 1850 held in the laboratory. J. Crustacean Biol. 26, 495–502.

Stevens, B., Armstrong, D., 1984. Distribution, abundance, and growth of juvenile Dungeness crabs, *Cancer magister*, in Grays Harbor estuary, Washington. Fish B-NOAA 82, 469–484.

Stevens, B., Armstrong, D., Cusimano, R., 1982. Feeding-habits of the Dungeness crab *Cancer magister* as determined by the index of relative importance. Mar. Biol. 72, 135–145.

Stone, R., O'clair, C., 2001. Seasonal movements and distribution of Dungeness crabs *Cancer magister* in a glacial southeastern Alaska estuary. Mar. Ecol. Prog. Ser. 214, 167–176.

Stone, R., O'clair, C., 2002. Behavior of female Dungeness crabs, *Cancer magister*, in a glacial southeast Alaska estuary: homing, brooding-site fidelity, seasonal movements, and habitat use. J. Crustacean Biol. 22, 481–492.

Strub, P., James, C., 1988. Atmospheric conditions during the spring and fall transitions in the coastal ocean off Western United States. J. Geophys. Res. Oceans 93, 15561–15584.

Sulkin, S., Blanco, A., Chang, J., Bryant, M., 1998a. Effects of limiting access to prey on development of first zoeal stage of the brachyuran crabs *Cancer magister* and *Hemigrapsus oregonensis*. Mar. Biol. 131, 515–521.

Sulkin, S., Lehto, J., Strom, S., Hutchinson, D., 1998b. Nutritional role of protists in the diet of first stage larvae of the Dungeness crab *Cancer magister*. Mar. Ecol. Prog. Ser. 169, 237–242.

Sulkin, S., Mckeen, G., 1989. Laboratory study of survival and duration of individual zoeal stages as a function of temperature in the brachyuran crab *Cancer magister*. Mar. Biol. 103, 31–37.

Sulkin, S., Mckeen, G., 1996. Larval development of the crab *Cancer magister* in temperature regimes simulating outer-coast and inland-water habitats. Mar. Biol. 127, 235–240.

Sulkin, S., Mojica, E., Mckeen, G., 1996. Elevated summer temperature effects on megalopal and early juvenile development in the Dungeness crab, *Cancer magister*. Can. J. Fish. Aquat. Sci. 53, 2076–2079.

Swiney, K., Shirley, T., 2001. Gonad development of southeastern Alaskan Dungeness crab, *Cancer magister*, under laboratory conditions. J. Crustacean Biol. 21, 897–904.

Swiney, K., Shirley, T., Taggart, S., O'clair, C., 2003. Dungeness crab, *Cancer magister*, do not extrude eggs annually in southeastern Alaska: an in situ study. J. Crustacean Biol. 23, 280–288.

Taggart, S., O'clair, C.E., Shirley, T.C., Mondragon, J., 2004. Estimating Dungeness crab (*Cancer magister*) abundance: crab pots and dive transects compared. Fish B-NOAA 102, 488–497.

Tasto, R., 1983. Juvenile Dungeness crab, *Cancer magister*, studies in the San Francisco bay area. Fish Bull. (Calif.) 1983, 135–154.

Tegelberg, H., 1970. Annual progress report: Dungeness crab study. Washington Department of Fisheries, 1–21.

Tegelberg, H., 1971. Condition, yield and handling mortality studies on Dungeness crabs during the 1969 and 1970 seasons. In: 23rd Annual Report of the Pacific Marine Commission for the Year 1970. Pacific Marine Commission, Portland, OR.

Thomson, R., 1981. Oceanography of the British Columbia coast. Can. Spec. Publ. Fish. Aquat. Sci. 56, 281.

Troffe, P.M., Levings, C.D., Piercey, G.B.E., Keong, V., 2005. Fishing gear effects and ecology of the sea whip (Halipteris willemoesi (Cnidaria: Octocorallia: Pennatulacea)) in British Columbia, Canada: preliminary observations. Aquat. Conserv. 15, 523–533.

Visser, E., Mcdonald, P., Armstrong, D., 2004. The impact of yellow shore crabs, *Hemigrapsus oregonensis*, on early benthic phase Dungeness crabs, *Cancer magister*, in intertidal oyster shell mitigation habitat. Estuaries 27, 699–715.

Waldron, K., 1958. The fishery and biology of the Dungeness crab (*Cancer magister* Dana) in Oregon waters. Oregon Fish Commun. 24, 1–43.

Wickham, D., 1978. A new species of *Carcinonemertes* nemertea *Carcinonemertidae* with notes on the genus from the Pacific coast. Proc. Biol. Soc. Wash. 91, 197–202.

Wickham, D., 1979a. *Carcinonemertes errans* and the fouling and mortality of eggs of the Dungeness crab, *Cancer magister*. J. Fish. Res. Board Can. 36, 1319–1324.

Wickham, D., 1979b. Predation by the nemertean *Carcinonemertes errans* on the eggs of the Dungeness crab *Cancer magister*. Mar. Biol. 55, 45–53.

Wickham, D., 1979c. The relationship between megalopae of the Dungeness crab, *Cancer magister*, and the hydroid *Velella velella*, and its influence on abundance estimates of *C. magister* megalopae. Can. J. Fish. Aquat. Sci. 65, 184–186.

Wickham, D., 1980. Aspects of the life history of *Carcinonemertes errans* (Nemertea: Carcinonemertidae), an egg predator of the crab *Cancer magister*. Biol. Bull. 159, 247–257.

Wicksten, M., 2009. Decapod Crustacea of the Californian and Oregonian Zoogeographic provinces. Scripps Inst Oceangr Lib Paper 26, 419.

Wild, P., 1983a. Comparison of ovary development in Dungeness crabs, *Cancer magister*, in Central and Northern California. Fish Bull. (Calif.) 172, 189–196.

Wild, P., 1983b. Effects of seawater temperature on spawning, egg development, and hatching success of the Dungeness crab, *Cancer magister*. Fish Bull. (Calif.) 172, 197–213.

Wild, P., 1983c. Life history, environment, and mariculture studies of the Dungeness crab, *Cancer magister*, with emphasis on the central California fishery resource. Fish Bull. (Calif.) 172, 1–352.

SUBJECT INDEX

Note: Page numbers followed by "*f*" indicate figures, and "*t*" indicate tables.

TAXONOMIC INDEX

Note: Page numbers followed by "*f*" indicate figures, and "*t*" indicate tables.